CHILDREN
of
the SUN

A History of Humanity's
Unappeasable Appetite for Energy

Alfred W. Crosby

W. W. Norton & Company

New York • *London*

For information about permission to reproduce selections from this book,
write to Permissions, W. W. Norton & Company, Inc.,
500 Fifth Avenue, New York, NY 10110

Manufacturing by The Haddon Craftsmen, Inc.
Book design by Brooke Koven
Production manager: Anna Oler

Library of Congress Cataloging-in-Publication Data

Crosby, Alfred W.
Children of the sun : a history of humanity's unappeasable appetite
for energy / Alfred W. Crosby. — 1st ed.
p. cm.
Includes bibliographical references and index.
ISBN 0-393-05935-9 (hardcover)
1. Power resources. 2. Power resources—History.
3. Solar energy—History.   I. Title.
TJ163.2.C78 2006
333.79—dc22

2005024361

W. W. Norton & Company, Inc., 500 Fifth Avenue, New York, N.Y. 10110
www.wwnorton.com

W. W. Norton & Company Ltd., Castle House,
75/76 Wells Street, London W1T 3QT

1 2 3 4 5 6 7 8 9 0

For Fran,
without whom, in my opinion,
the sun would expire.

CONTENTS

LIST OF ILLUSTRATIONS

Ancient religious institutions that considered terrestrial creatures, especially man, to be children of the sun were far nearer
the truth than is thought by those who see earthly beings as
ephemeral creations arising from blind and accidental interplay
of matter and forces.

—VLADIMIR I. VERNADSKY, GEOCHEMIST (1926)

The force that through the green fuse drives the flower
Drives my green age; that blasts the roots of trees
Is my destroyer.

—DYLAN THOMAS, POET (1934)

PREFACE

Children of the Sun is a history of how we access the energy to get work done, to move our muscles, to think, to hunt mammoths, to sow and harvest, to build pyramids, to power automobiles and space rockets, to boil water for tea. Very nearly all of the energy needed to do these things originates in the core of the Sun and travels as sunlight to Earth, where photosynthetic plants process it into leaves, stems, flowers, seeds, and roots. Animals like us eat these plants and/or animals that feed on this biomass and thus gain the calories to function.

Our next step toward biosphere domination (or wherever we're headed) was to discover how to tap sun energy concentrated in biomass—in wood, for instance—by burning it to produce light and heat where and when we wanted. We became the only species to control fire and to direct it for our purposes. Our ability to cook food—to predigest what we couldn't or preferred not to digest raw—gave us more efficient access to greater stores of nutrients. Our numbers increased. The next step was agriculture: with cultivated plants and livestock we assured ourselves of greater supplies of energy than we had ever had before. We increased again in numbers and geographical expanse.

And so we sat for our first few thousand generations, by half of which we were the most widely distributed of all large animals in the Eastern Hemisphere, and by no less than 10,000 years ago had achieved that status in all continents but Antarctica.

We were a successful species, but by our current standards nearly helpless. Our primary means to get work done was muscle, our own and animal. Yes, by AD 1000 we had windmills, watermills, and sailing ships, but in truth our societies ran on muscle, as illustrated by any number of old pictures of our ancestors slogging along on treadmills to provide the power to grind grain, lift water, and so on. We had gone about as far as we could go in numbers and altering the world in accordance with our wishes unless we utilized more efficient means of tapping sun power than agriculture and burning wood.

We lurched into the fossil fuel era some two to three hundred years ago with the invention of the steam engine. Like its successor, the internal combustion engine, the steam engine enables us to tap the concentrated energies of ancient biomass which subterranean heat and pressure have transformed into coal, oil, and natural gas. We created the means to transmit the energies we harvest by burning fossil fuels hundreds, even thousands of miles, by expressing it as electricity. Our technological civilization as it now exists would be impossible without the enormous consumption of these fossil fuels. Modern civilization is the product of an energy binge.

Binges often end in hangovers. Fossil fuel supplies are ultimately exhaustible and currently responsible for such worrisome effects as global warming. We must revive old ways of tapping sun power, such as windmills, and invent new ways to do so, such as solar cells, and/or we must utilize new sources of energy. Nuclear fission could produce all that we need, but is distrusted as dangerous. It may be Mother Nature's version of the Trojan Horse. The ideal solution may well be hydrogen fusion, the Sun's means of producing energy; but we don't know how to recreate that process as a practical procedure and may never learn how.

We have successfully met the energy challenge several times before with breakthroughs such as the invention of agriculture and the steam engine. But humankind's unappeasable appetite for energy makes the solutions ephemeral, and the challenge permanent.

*   *   *

BEFORE I INVITE you to proceed to a more detailed consideration of these matters, I would like to thank the many who read all or parts of my manuscript and offered corrections and suggestions (which I usually but not always accepted, so their responsibilities are limited). These include Frances Karttunen, John McNeill, Steven Stoft, Kurt Keydel, Vladimir Strelnitski, Jerry Bentley, and my patient editors, Steve Forman and Sarah England.

PART ONE

THE LARGESS OF THE SUN

The Sun . . . contains 99.86 per cent of the mass of the solar system.

—*John Gribbin, cosmologist (1998)*[1]

I do good to all the world. I give them light and brightness that they may see and go about their business; I warm them when they are cold; and I grow their pastures and crops, and bring fruit to their trees, and multiply their flocks. I bring rain and cold weather in turn, and I take care to go round the world once a day to observe the wants that exist in the world and fill and supply them as the sustainer and benefactor of men.

—*Garcilaso de la Vega, El Inca (1612)*[2]

Great scientists, great artists, great athletes—Einstein, chalk in hand at the blackboard; Michelangelo, paint dripping down his arm, working on the ceiling of the Sistine Chapel; Lance Armstrong wheeling across the finish line of the Tour de France—strike us as wellsprings of energy. But no human being, indeed no life form whatever, produces more energy than it takes in, or produces any at all by itself. All humans and all organisms are dependent on external sources for fuels. All are parasitic.

[1]John Gribbin, *Almost Everyone's Guide to Science: The Universe, Life, and Everything* (New Haven: Yale University Press, 1999), 169.

[2]Garcilaso de la Vega, El Inca, *Royal Commentaries of the Incas and General History of Peru*, trans. Harold V. Livermore (Austin: University of Texas Press, 1966), 42–43.

The two principal sources of energy upon which we earthlings depend are the upwellings of heat, magma, and gases from within the planet and the radiation from the Sun. The former empowers exotic organisms such as those that live by the hydrothermal vents of the deep ocean; they won't be mentioned again in this book because they have little direct influence on human history. The latter source, sunlight, is by every measure the greatest source of energy, the fuel of life, on the surface of our planet. Here we are all, in the words of Vladimir Vernadsky, the Russian geochemist, "children of the sun."[3]

The Sun is the center of a dust twirl of planets and lesser specks, and is also the center of human life and of the life forms on which humans are dependent. And no wonder: that star makes up very nearly 100 percent of the mass of the solar system. Our Earth is no more than a mote of debris left over from its formation. A million of our planet would fit inside the Sun.

At its core, where the environment is exponentially many times more hellish than Dante could have conceived, the pressure and temperature are so extreme that no solids, liquids, or gases can exist, only plasma, the uninhibited swarming of subatomic particles which is the fourth possible state of matter. There, nuclei (the relatively massive central points of atoms) of hydrogen undergo a transmutation that is the foundation of our lives. Despite being of like and mutually repellent electric charge (like similar poles of magnets repelling each other, north versus north, south versus south), they collide head-on because there is no way of not doing so. When four hydrogen nuclei collide and fuse, the result is one helium nucleus.

The mass of the one helium nucleus is 0.7 percent less than that of the four hydrogen nuclei, and the missing mass converts into pure energy. That is, per collision, a very tiny amount, but there are trillions of trillions of these collisions per second. Furthermore, the quantity of mass involved is not as decisively important as what

[3]Quoted in Vaclav Smil, *The Earth's Biosphere: Evolution, Dynamics, and Change* (Cambridge, MA: MIT Press, 2002), 9.

happens to it. To measure that, we resort to Albert Einstein's famous formula for calculating the quantity of energy represented by a given mass, $E = mc^2$ (Energy equals the mass times the constant squared). Said mass may be small, but c, the constant, is the speed of light—*squared*. This is the awesome conversion which we replicate on a minuscule scale with our hydrogen bombs. It is no wonder that we are advised not to look directly at the Sun.

The energy created by hydrogen fusion in the Sun's core rises to its surface and blows out into space in all directions as light, the fuel of life. Approximately eight minutes later and 93 million miles away, the mote which is our planet receives a half billionth of this radiation. Half of that is reflected back into space or absorbed by atmosphere and clouds. The paltry remainder is the largess that made and makes life on our planet's surface possible, including the lives of Einstein, Michelangelo, and Lance Armstrong.

Life on Earth began an immensity of time ago and for millions upon millions of years thereafter was limited in its greatest extravagance to prokaryotes, one-celled organisms without nuclei. One of the most common of these was cyanobacteria, blue-green algae. They made their living by directly tapping sunlight. Our name for their chemical masterstroke is photosynthesis (from Greek words meaning "light" and "to put together"). Cyanobacteria contain a greenish substance, chlorophyll, which makes the alchemists' hyperbolized Philosopher's Stone seem feeble. Chlorophyll absorbs and harnesses the energy of light to split water and carbon dioxide molecules. This makes other molecules, simple carbohydrates, the fuel from which and by which yet other molecules essential to the functions of life are constructed.

At an inexpressibly important moment long, long ago certain enterprising prokaryotes or perhaps new-fangled eukaryotes (cells with nuclei) engulfed cyanobacteria and did not digest but incorporated and recruited them. These cyanobacteria lost all ability to exist independently and settled down forever to sinecures as distinct entities called chloroplasts inside their hosts. They are there still, almost all of their needs filled by their landlords, who in the aggregate are our plants. In return for lodging, the chloroplasts absorb sunlight (they will even migrate within cells to gain better access to the light) in order to produce and supply their hosts with simple sugar, the basic food of both plants and animals. Human food chains always ultimately lead down to plants with chloroplasts arranging themselves to catch the sun's rays.

In the process of photosynthesis the oxygen component of the dismembered water molecules is discarded and drifts away. This is the source of most of the oxygen in our atmosphere, which animals like us take up and, via slow combustion (the technical term is "respiration"), turn the food we have ingested into energy. We are thereby able to function, to build and rebuild ourselves, to stay alive from minute to minute. The ash of this sedate burning is carbon dioxide, which we exhale, repaying the biosphere for the carbon dioxide which photosynthesis has used up. The energy that respiration produces drives our muscle to do "work." In physics, that noun specifically refers to the transfer of energy from one physical system to another, as when I apply pressure to these keys to type this sentence.

We gain access to that force via "prime movers" or indirectly from entities that tap prime movers. The prime mover concept dates back to Aristotle, and to St. Thomas Aquinas, who posited a mover who is never moved, but moves everything else, i.e., God. Physicists have pared the concept down to mean any machine (using that word broadly) that converts natural energy into work. I utilize muscle, humanity's first prime mover (it taps the oxidation of carbohydrates, a natural force), to press down my computer keys.[4] The windmill, a later prime mover, taps the movement of air to do the work of turning millstones and grinding grain into flour. A standard nuclear reactor taps the heat produced by atomic fission, a nat-

---

[4]Steven Vogel, *Prime Mover, A Natural History of Muscle* (New York: W. W. Norton & Company, 2001), is a good place to start on this subject.

ural process, to transform liquid water into steam to drive turbines to provide us with electricity. (We use electricity to drive locomotives, elevators, streetcars, movie projectors, computers, and so on. You could think of them as secondary movers if there were such a category.)

When we first started, we fueled our personal prime movers with the food we acquired as simple harvesters of wild plants and animals. After a long while we elevated ourselves to the status of complicated harvesters. We domesticated (negotiated alliances with) a few other species, as have, for instance, ants (as we shall see in chapter 3), so we could have sources of food, hide, fiber, bone, and help within easy reach. We harnessed fire (as have no other creatures), tapping sun energy by igniting biomass created by photosynthesis. The burning of recently living biomass—wood, for instance—has continued on into our time. In the last two centuries we have also been burning immense, almost immeasurable, quantities of fossilized biomass from ages long before our species appeared. Today, as ever, we couldn't be more creatures of the sun if we went about with solar panels on our backs.

In the last half century our demand for energy has accelerated to the verge of exceeding what is produced and can be produced by conventional ways of harvesting sun source energy. We are refining those ways, which typically tap the energy holding molecules together, through combustion, for instance. We are also trying to domesticate the energy holding atomic nuclei together, such as is extravagantly released in fission and fusion bombs. Our sun-struck physicists are even leapfrogging back over photosynthesis and committing hydrogen fusion in their laboratories.

The above, as we shall see in more detail, can all be counted as triumphs in the quest of the children of the sun for more energy. We will also see some failures in that quest. As a historian of large-scale change, I sift through truckloads of books and journals and try to come up with generalities to save my readers from being overwhelmed by what may seem to be temporal and spatial chaos. But I worry about sounding like a high school teacher of mine who said there were four causes for the French Revolution, and when a student suggested a fifth, said no, there were only four. As an amulet against oversimplification, at the end of most of the following chapters I will add a coda about a person or event with the texture and grain of specificity (and occasionally with something that may even contradict my most recent pontifical pronouncement). In this way I hope to counter generalities that, like sleek limousines sweeping tourists from airports through gritty city neighborhoods to comfortable hotels, deprive readers of a sense of the ambiguities and the seemingly irreconcilable details of our past.

I

FIRE AND COOKING

During the flood, the monitor lizard and his wife, the civet cat, managed to save fire by carrying it up a tree. Those ancestors who could not climb the tree became fish and other forms of sea life. Once the water had subsided, those ancestors who could not climb down the tree became birds. Only the monitor lizard and the civet cat . . . were able to keep the fire in a clay pot. They would scare all the other animals away or burn them to death with this fire.

—Legend of the Onge of the Andaman Islands[1]

The Earth is unique among the planets of our solar system in its inclination to catch fire. Other planets and some satellites have their lightning storms and volcanoes erupting streams of red-hot lava, but they don't have fire. Lightning and lava can set trees and houses aflame, but they do not burn in and of themselves. Lightning is electricity, not fire, and rock does not burn. Burning is a specific chemical process and few substances in the universe can burn. Organic matter—wood, grass, manure, your hair—can. Photosynthesis makes this biomass plentiful here on Earth and laves our atmosphere with oxygen, assuring that ignition and fires are possible, indeed unavoidable.

Oxygen, the leftover of photosynthesis, is highly reactive, that is to say, it combines molecularly with a great variety of elements.

[1]Madhusree Mukerjee, *The Land of Naked People: Encounters with Stone Age Islanders* (Boston: Houghton Mifflin, 2003), 235.

Indeed, it forms compounds with practically all chemical elements except inert gases, and hence is the most common element in the Earth's crust. The process of combination, called oxidation, is accomplished at a range of rates. Digestion, and the rusting of iron, are slow kinds of oxidation. Burning is fast, and the explosion of gunpowder is spectacularly fast, oxidation.

Our planet is also unique in having inhabitants to ignite and control, to *domesticate,* fire: us. When our ancestors learned to make knives and axes of stone, they were only producing extensions of their puny teeth and fingernails. In contrast, when they learned to manipulate fire, they were doing something truly unprecedented. Other apes, orangutans, for instance, have in our day toyed with hot coals and flame, but that is little more than a trick, like a dog walking on two legs. Only humans have had the courage and manual dexterity to develop a technology, or one could say, a culture of fire. Charles Darwin considered that advance, that seizure of a new, swift, and reliable means to tap the sun energy accumulated in organic matter, as more important in humanity's history than any other, excepting the development of language.

Fire lighted our ancestors' way at night and deep into caves. Fire helped our ancestors to fend off predators. Fire allowed them to survive winter nights out of the tropics. Fire equipped them to harden wooden spearpoints and to drive game, no matter how huge and dangerous. Fire enabled our ancestors to burn off forests to make grasslands in which animals to be preyed upon and animals to be feared could be easily seen. Fire empowered our ancestors to burn off old brush, dead straw, and leaves to make way for tender green plants to exploit for a myriad of purposes, such as attracting delectable animals like deer. Fire armed our ancestors with imperial power over flora and fauna, over whole landscapes, such as no other creatures had ever wielded before.

The greatest advantage gained by domesticating fire may have been—no, let me say flatly, *was*—in a practice we usually do not dignify as world-shaking. I refer to the salubrious effect of high heat on our food and therefore on our bodies. I refer to cooking.

Our most instructive artifact from the primordial history of our diet is our bodies. We have inherited them from the hunter-gatherers of the first 90-plus percent of our existence as the distinct species we modestly call *Homo sapiens*, which in Latin means "wise man." Compared to our cousins, the chimpanzees and other apes, and the members of our immediate family and direct ancestors, the hominids, we have, in proportion to our body size today, very big brains, small teeth and associated bone structure and musculature, and small guts (gastrointestinal tracts).

The snap explanation for our big brains is that we evolved not to be physically stronger or faster or fiercer than our predators and prey, but smarter. One wonders why we started down that path. Intelligence has performed wonders for us, but that certainly wasn't obvious a few million years back. A few points of IQ gained per millennium wouldn't seem to have been all that valuable. A step faster to the nearest tree when the lions showed up would have been of greater immediate advantage than an elaboration of chimpish smarts, and evolution is the product of short-term effects. The explanation for the distinctively smaller teeth and gut is even more elusive.

To shed some light on the origins of these, we go back to the evolution of food procurement prior to the invention of cooking. Humans are born with four basic tastes: sweet, sour, bitter, and salty. A clear majority of us prefer sweet, and not just because we are bombarded by ads for candy bars and soft drinks. Babies, with no prompting from anything but their own taste buds, prefer sweet. It seems likely that hominids have had a sweet tooth for millions of years and for good reason. Sugar converts into energy faster than other foods. You have to dodge the lion's leap in the short run in order to qualify for the long run.

In the tropics where the early hominids lived, fruits were, as they are now, the most common sources of sugars. Three quarters of a chimpanzee diet consists of fruit, and so it was, we can assume, for the early hominids. Fruit trees are not uniformly distributed across tropical woodlands. A foliage-eating gorilla can start nearly anywhere in its home forest and eat till full, but a fruit-fancier's next meal, even survival, depends on searching for, locating, and then remembering where the fruit trees are. It is no coincidence that primates that depend on fruits generally have bigger brains relative to their size than the others.

Sugar and the other carbohydrates supply the fuel to drive the engine of our bodies but not to build and maintain it. For that proteins, lipids, minerals, and vitamins are needed. Fruits supply some of these, but not all and not enough. The early hominids, like present-day chimps, must have obtained a good portion of theirs by consuming delectable insects and occasional meaty treats that scavenging and ambushing smaller animals might produce, but chiefly they did so by feeding on lots of foliage and other parts of plants. We've been eating salads for millions of years.

Fruit trees hospitably surround their seeds with sweet flesh in order to attract the consumer, who will deposit the seeds, along with fertilizer, away from the mother plant. Other plants have other strategies. Many are inhospitable to browsers in hopes of surviving long enough to produce and distribute seed by wind, by sticking to birds' feet, and other means. For humans, a lot of plant matter ranges from the hard to digest to the impossible to digest to the poisonous. It contains, for instance, lots of fiber nearly as defiant of teeth and gastric juices as plastic, plus various toxins to discourage consumers by tasting bad and by making them sick. (Many humans, oddly enough, quite fancy several of these poisons—caffeine, nicotine, morphine, cocaine.)

Our motherland, the African rain forest and savanna, is not and never was a salad bar. The early hominid plant foods must have been limited to fruits, seeds, stems, and leaves (preferably immature and tender because these are low in fibrous cellulose and toxins) and underground storage organs: bulbs, tubers, and corms. Those of this last category, especially the ones deepest underground, may have been a vital source of nourishment during famines, which functioned as genetic bottlenecks because only the survivors propagated. It is likely that the hominid's first epochally important tool was not the stone ax but the wooden digging stick.

Such a diet enabled a number of species of apes to live and propagate in their ancestral homelands and to migrate to contiguous and climatically similar regions. But these foods could never have enabled hominids to migrate across the Old World tropics, nor their only surviving species, *Homo sapiens*, to occupy all the ice-free continents and most of the larger islands of the whole planet. The chimp, with its big teeth for tearing and chewing leaves and big gut for digestion of their coarse materials, gets by on such a diet, but could never have moved to and prospered in the Scandinavian forest. Apish diets limited early hominids to being no more than another parochial kind of ape.

In order to adapt to different environments—deserts, temperate zone forests, tundra, each with its distinctive plant and animal food sources—hominids either had to wait many, many generations for their bodies to adapt genetically (with adaptations peculiar to each different environment) or they had to do something clever. They would have to devise methods to broaden exploitation of the organic yield of the Sun's largess, a search that led, eventually, to cooking.

Hominids probably started by scavenging more effectively, that is, learning how to scare other carnivores and omnivores away from the delectable dead. The ability to throw hard and accurately, a unique trait of *Homo sapiens* today that immediately ancestral species may have possessed, would have helped with that. They may have improved their hunting skills, graduating from morsels like monkeys to meal-sized animals like gazelles. Meat calorically is two to four times richer than fruit, and provides a dense package of fats and proteins already processed for swift digestion and incorporation as muscle, bone, and other organs. Meat includes the amino acids that the human body cannot assemble for itself, cannot store, and can obtain from flora only by dining on many different kinds of plants. The availability of animals to eat in periods of plant scarcity—winters, droughts, and such—and in regions such as deserts and tundra where plant food may be scarce is another advantage for hunting.

The adoption of hunting accelerated the trend toward commu-nality, toward family, clan, and tribe. Whether chimps or hominids, a group on the hunt for meat had a much greater chance of success than a single individual. But hunting, and scavenging, too, involves risk; it is not a trip to the grocery store. A good day may win you a week's nourishment; a bad fortnight may leave you exhausted and starving. One mistake could get you killed. It is not surprising that meat comprises only a small percentage of chimpanzee diets and often less than half of hunter-gatherer diets today.

Hominids did not bet their survival on hunting exclusively or, one would guess, even primarily. They invented a new way to tap stored photosynthetic energy, a way even more innovative than birds' and bats' mastery of flight. Feathers and wings are products of mindless evolution. Cooking is cultural, not genetic, an unprece-dented innovation.

Our ancestors chose to increase the proportion of available organic matter that they could digest by inventing means to per-form some of the essential processes of digestion outside the body—by cooking. Treating food, plant or animal, with high heat changes it, simplifies it, so to speak, so our teeth and gut can deal with it more effectively. Cooking tenderizes hide and husk; pops open cells to digestive juices, changes protein and starch molecules so our enzymes can get at them, makes small digestible molecules out of big indigestible molecules, kills bacteria, and defangs many toxins. In general it transforms organic matter, which when raw is unpleas-ant to eat, difficult or impossible to digest, unhealthy, and even deadly, into nourishing, palatable food. Cooks didn't invent new prime movers, but they gave us a source of new fuel for the existing prime mover: muscle.

Take, for instance, the cereal grains that are today the staples of humanity. We domesticated small-kerneled grains like rice, wheat, barley, and rye not only for their qualities as nourishment, but also for their resistance to spoilage when stored for long periods. But eating such durable grains requires more than just popping them into your mouth. Imagine yourself lost with nothing but a handful of rice (kernels as hard as pebbles) to eat. To survive you might need water, and fire, to boil them.

Cooking is universal among our species. No explorer ever found a human society that did not cook. Cooking is more unequivocally characteristic of our species than language. Animals do at least bark, roar, chirp, sending signals by sound; only we bake, roast, and fry. Cooking also enormously increased the range of organic matter that hominids could tap for nourishment, and thereby the kinds of places and climates in which people could live. It seems likely that our immediate ancestral species, *Homo erectus* (1.7 million to 200,000 BP, years before the present), who migrated out of Africa and across Eurasia's lower latitudes, had fire as an aide. It is certain that *Homo sapiens*, which by a dozen millennia ago was living in all the continents except Antarctica and even in the loom of continental glaciers, had fire.

The tending of fires and cooking probably were female specialties from the beginning. The first step in these practices is the gathering of kindling, and women were the gatherers of the hunter-gatherer team. They collected biomass to burn, along with plant foods, and carried them to a communal center where the kin, the group, came together around the fire to eat. Feeding before had always been done at the spot when and where the food was obtained. Chimps do not delay consumption to carry food back home to the family. We do. The Harvard anthropologist Richard Wrangham proposes that humanity was launched when a kind of ape learned how to cook. In his view we are not herbivores or carnivores, but "cookivores."

Cooking may in fact have altered our bodies. Our brains tripled in size over the last 4 million years, a rapid change by evolution's standards. Cooking, like hunting, obliged human hunters, gatherers, fire tenders, and cooks to plan and cooperate—*to think*—and this may have helped to drive the transformation. Cooking may have gained for us not only enough calories, but enough leisure time to exercise our new brains gainfully. Chimps spend six hours a day chewing; cookivores only one.

We can now conclude that the decrease in the size of our teeth and gut is at least in part the result of cooking. The domestication of animals has usually featured a decrease in ferocity and in the size of claw and tooth. Although we have no way to directly measure the evolution of hominid ferocity, we have apparently always had fingernails, not claws. Our ancestors had and we still have sharp teeth and the jaws and muscle to operate them, but these have diminished relative to body size. Five million years ago, we probably could crack bones for their marrow with our teeth. Today, with our puny masticatory assemblage, we can barely manage celery. We would find it hard to eat enough to stay alive without softened, cooked food.

The history of our gastrointestinal tract is an elusive subject, but it does seem linked to diet change. Two of the most metabolically expensive organs in the human body (i.e., the greediest consumers of photosynthetic yields) are the brain and the gut. No animal has, relative to body size, both a very big brain and a very big gut. Cows evolved to have brains that are less impressive than their stomachs and intestines, which in their tunnels and chambers digest what we would find as indigestible as wood. Chimps evolved to a different balance between brain and gut. They have big brains compared to cows: the brains of chimps at rest consume 8 to 10 percent of their energy budget. Chimps have smaller guts than cows, though big enough to meet the challenge of a lot of coarse plant matter. Hominids, at least those directly ancestral to us, put most of their money on getting smarter. The ultimate result was *Homo sapiens*, a creature with an enormous brain that consumes 20 to 25 percent of its energy budget and with teeth too small and a gut too slight to cope with a chimpanzee's diet or, presumably, the earliest hominid diet. The sum of the metabolically expensive organs of a human and an ape of equal size are approximately equal in size, but the proportions of brain to gut are very different and just what you would expect: the ape has the bigger gut and the human the bigger brain. So cooking may well have decisively influenced our gut as well as our brain.

The question of how long cooking has been going on is a very difficult one. We can't even be sure when hominids adopted fire as a servant. Millions of years ago? Tens of thousands of years ago? And what do we mean by cooking? To scorch the legs and wings off delectable grasshoppers is simple, but is it cooking? You can bake food by burying it in hot ashes, but the food gets dirty. You can boil food by dropping it and hot rocks into a pit filled with water, but it might be better to invent some sort of pot. Millennia may have passed between scorching off grasshopper appendages and the preparation of oatmeal mush, much less pâté de foie gras. There are a few reddish patches at Koobi Fora and Swartkrans in Africa a million and a half years old that may be the remains of hominid fires— or, on the other hand, may be evidence of slowly burning tree trunks or lightning strikes or what-have-you. Advances in technology may one day allow us to identify these as hominid in origin, but for now we are stymied as to the date when our ancestors first domesticated fire.

What do I mean by "domesticated"? Fire must at first have inspired awe and fear more than an eagerness to use it. Domestication of fire did not begin with ignition by our own means but by obtaining fire as burning branches or hot coals from conflagrations ignited by natural means such as lightning. (Right down to modern times there have been a few remote peoples, some Andaman Islanders and Congo Pygmies, for instance, who did not know how to start a fire but only how to preserve it as coals or smoldering chunks of biomass.) Fire, immortal if fed and beyond revival if neglected too long, must have seemed sacred rather than efficacious.

Candles still burn on our altars for our most important ceremonies.

Haim Ofek, the anthro-economist, suggests that fire was so precious to our early ancestors that it was probably one of the earliest items we bought and sold. There were so many advantages to having one's own fire that its dying out was a real crisis and reignition a necessity. A keeper of a flame lost nothing by giving flame away (it is not matter per se but an easily extendable process) and could require valuables and services in return.

Eventually our ancestors invented ways to make fire whenever they wanted by rubbing sticks together and by striking sparks from flint and the like. With that, our reliance on flame and glowing coal became irreversible. The date of that breakthrough must have been long after our ancestors first picked a burning branch from the hot ashes of a natural fire and shook it to see the sparks fly. But how will we ever find out the date for either breakthrough?

We may have more success by considering cooking and asking when it became standard practice and, therefore, undeniably influential? *Homo erectus*, the first hominid we would recognize at a glance as human, appeared nearly 2 million years ago. Its brain was bigger and its teeth and gut smaller than earlier hominids. These details suggest a diet including lots of meat and/or cooked foods. Was that creature so efficient a scavenger and hunter as to have a diet primarily of flesh? Today's tropical hunter-gatherers in places like Africa, Asia, and New Guinea, with far bigger brains, often fail to manage that. So *Homo erectus* must have cooked so as to extract nourishment from a broad selection of organic matter.

The only thing that makes us hesitate to accept that conclusion is the weak link in many a brilliant theory—the lack of evidence. To credit *Homo erectus* as a chef we need, oh, a dozen 500,000-year-old sites of baked soil or rock in conjunction with burned animal bones and remains of stone tools. We don't have them, perhaps because *Homo erectus* lived in the tropics and did not take shelter in caves, where evidence of fires is more apt to last than in the open. Or perhaps *Homo erectus* was too mobile to build fires in the same spot often enough to leave strong traces. Be all that as it may, the best proof we have for *Homo erectus*'s mastery of fire is a matter of reasonable inference, which, however appealing, is a thin reed to lean on. We cannot be sure that cooking figured significantly in the evolution of hominids before *Homo sapiens.*

That species, ours, appeared quite recently, only 150,000 to 200,000 years ago, but early *Homo sapiens* bones are not commonly associated with clear evidence of either camp fires or cooking. They and/or the Neanderthals[2] had hearths, but these were little more than surfaces on which fires had burned. Then, about forty or fifty thousand years ago, give or take ten thousand years, lots of hearths began to appear, some of them with stone liners and ditches to enhance air intake and other elaborate features. Hearths in quantity are one of the diagnostic characteristics of the Upper Paleolithic, a pivot point, perhaps *the* pivot point in our history. (The Upper Paleolithic is usually dated as 40,000 to 10,000 or so BP, but the numbers vary from region to region and change as new data comes in.)

Let me note here that the great majority of these hearths have been found in Europe not because Paleo-Europeans were necessarily the first or best fire makers, but probably because most Old World paleoanthropologists have been European and their home continent is where they have done most of their field research. Recent discoveries and general professional experience indicate that earlier examples will be found in Africa, where most of the innovations in the early history of *Homo sapiens* made their debuts. Meanwhile, we have to make do with European examples.

If there had never been an Upper Paleolithic, we would still have to credit the *Homo sapiens* species as remarkable. On the eve of the Upper Paleolithic, humans were scattered in the Old World (East-

---

[2]I have purposely omitted the Neanderthals because they probably contributed little to the advance of material innovation compared to *Homo sapiens*, the species that replaced or absorbed them.

ern Hemisphere) from Atlantic to Pacific, from near the glaciers south to the oceans. They were already in or were about to embark to Australia, which required a sea voyage of about 90 kilometers. The number of different and identifiable tools they made and used was twice the total of those their predecessors possessed and, we are probably safe in assuming, their behaviors were more various and complex, as well. They were, nonetheless, staid by the standards of their successors. Their bodies were like ours, but their cultures much simpler. Their tool kits included no more than forty or so items and the designs of these were uniform over enormous areas and changed hardly at all for thousands of years. They left few indications of focused religious belief and very little in the way of artistic tendency.

Then something triggered the acceleration which has continued until our time and led us to our present triumph and plight. Hominid bodies and behaviors had evolved in crude synchrony before. Now bodies stopped evolving and cultures raced ahead.

The kinds of tools Upper Paleolithic humans made and used soared in number to perhaps a hundred. They developed new chipping and flaking skills in shaping stone, which produced more cutting edge per effort than ever before. Thin blades twice as long as wide and with keen edges became standard. Humans began to exploit materials seldom used previously: bone, horn, and antler. They invented awls, fishhooks, and eyed needles. They invented the atlatl (throwing stick) and, eventually, the bow and arrow, making themselves nature's most formidable hunters and each other's worst enemies. In barely more than twenty thousand years in the south of what is today France, they plunged headlong through four traditions of stone-bone-antler craftsmanship, each characterized by unique artifacts. Roger Lewin of Harvard's Peabody Museum calls that pace of change "almost hectic."[3]

[3]Roger Lewin, *Human Evolution: A Core Textbook* (Malden, MA: Blackwell Science, 1998), 431, 436.

Humans began to covet and seek afar for materials worthy of their new skills. In Upper Paleolithic sites archeologists have found high-grade flint, amber, and seashells tens of miles, even a hundred miles, away from their points of origin. Humans had added commerce and/or long-range raiding to their repertoire of behaviors.

They invented art: they painted and incised decorations on their tools that improved usefulness not at all, but rendered them somehow more pleasing. Expert shapers of stone produced masterpieces of no utility whatsoever, blades that are 20 centimeters long and so thin and delicate that they are translucent. Designs changed and kept changing across geographic expanse and time: style seems to have arrived. Humans began to shape clay statues of animals and humans, the grossly sexual and famous Venus figurines, for instance.

These humans took to living in a variety of shelters from caves to houses, constructed not just of logs and thatch but of stone blocks.

Venus of Willendorf, limestone figure from the Upper Paleolithic, 25th mill. BCE.

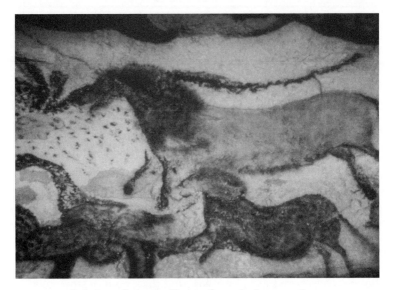

Painting of a herd of horses from the Lascaux Caves
in Périgord, Dordogne, France.

In steppe country such as Ukraine, with few caves or other natural
shelters to offer protection from the elements, humans had to
become builders. They made use of the materials at hand. There on
the treeless plains they built their homes of mammoth bones.

Their burial of valuables with the dead suggests that humans
began to believe in existences beyond the immediate and visible. At
Mal'ta, Siberia, archeologists have opened the 15,000-year-old grave
of a child launched on his or her way to eternity equipped with a
diadem, a necklace, a bone button, a bird figurine, a bone point,
and several blades. Symbolic thought and religious belief were
already in residence and probably had been for some time.

The greatest bequest of the Upper Paleolithic for most of us is
the cave paintings of Franco-Cantabria depicting wild animals long
gone from the region and some even from the world: bison, horses,
reindeer, mammoths, woolly rhinoceroses. At least eleven thou-

sand years ago, and in many cases much longer ago than that, for reasons we will never know, humans of stunning talent, devoting their time and energy to their art while others provided for them, crawled far back into caves and, by the light of crude oil lamps created masterpieces on the rock walls hundreds of meters from the sunlight.

In the Upper Paleolithic, humans attained unprecedented levels of skill in a variety of crafts and arts, adopted new habits of perception and thought, and, we must assume, achieved greater complexities of social organization than ever before. One scientist has proclaimed this as "the most dramatic behavioral shift that archeologists will ever detect."[4]

If the Upper Paleolithic marked the opening chapter of our transition from being very clever apes to the keystone species of the planet, the trigger for that shift is perhaps the deepest of the mysteries about our past. Our speculations about this key development are biased by our assumption that the cause must be worthy of the effect, which is, after all, nothing less than—*us!* A favorite candidate is the burgeoning of language. That is certainly a cause worthy of the effect, but there isn't and cannot be any direct evidence of speech as such in the Upper Paleolithic or thereafter until the advent of writing.

The best available physical evidence pertaining to the origins of speech is in the shapes of the skulls and neck bones of possible speakers of the past. Humans of the Lower Paleolithic seem to have been anatomically as well equipped for speaking as we are. Whether they had full languages or some primitive form thereof, we cannot know. The evidence is inadequate for anything but misty supposition. The rule of the scientist and scholar is to put unprovable hypotheses on the shelf and go on to consider others for which there is some evidence.

[4]Richard Klein, quoted in Haim Ofek, *Second Nature: Economic Origins of Human Evolution* (Cambridge: Cambridge University Press, 2001), 179.

LET US CONSIDER cooking, though, as said before, it be inglorious. In the Upper Paleolithic the evidence of hearths and cooking becomes, for the first time, plentiful and unambiguous. The dissemination of the practice of cooking throughout the species should have led to a healthier and, therefore, increasing population. The number of sites of human presence surviving from the Upper Paleolithic is greater than from earlier times, strong evidence that the population expanded. Cooking surely contributed to that. More people meant more trade in artifacts and ideas; more experimentation and invention; and more stimulation for speculating about the natural and supernatural.

Cooking required members of a band to gather at a single location to eat, and thus it multiplied the socializing influence of the fire. (We are still sensitive to its hypnotic influence, hence the appeal of fireplaces during holidays, of campfires during vacations.) It was around the fire gnawing on barbecued ribs that humans supplemented kinship with friendship and the band exfoliated into the tribe. Andaman Islanders of the Indian Ocean, who were in material culture so, shall we say, Middle Paleolithic that they could not make fire but had to carry it with them, afforded the anthropologist Alfred R. Radcliffe a glimpse of that effect a century ago. He wrote that for each islander, the individual life was only a "fragment" of "the social well-being which is the cause of his own happiness, [which] depends on the possession of fire. . . . Indeed, the Andamanese believe it is the possession of fire that makes human beings what they are and distinguishes them from animals."[5]

Could it be that Eve altered the course of history not by picking the apple, but by baking the first apple pie?

[5]Quoted in J. D. Clark and J. W. K. Harris, "Fire and Its Roles in Early Hominid Lifeways," *African Archaeological Review*, 3 (1985), 22.

Coda: Charles Darwin and
the Chefs of Tierra del Fuego

In December 1832 Darwin, supercargo on HMS *Beagle*, visited Tierra del Fuego, where, he wrote, "man exists in a lower state of improvement than in any other part of the world." He was referring to the Fuegans, aquatic nomads who often went naked in the rain and sleet of the southern tip of South America. The males wore animal skins about as large as a handkerchief on their backs, which they moved from side to side as the wind shifted. For shelters they threw together huts of branches and grass, the work of no more than an hour.

The Fuegans lived on seafood: shellfish, which the females gathered from the rocks and dived for, and finned fish, which they caught without hooks. The men hunted seals, sea lions, and such. Dead stranded whales were a special treat. The thoughtful often buried whale blubber for use later as famine food.

Darwin tells of a party of 150 or so Fuegans prevented by continual gales from gathering shellfish or hunting. Starvation loomed, so four men went out into the storm for food. They returned "carrying a great square piece of putrid whales-blubber with a hole in the middle, through which they put their heads, like the Gauchos do through their ponchos or cloaks."[6] Food of a sort had arrived, but it was marginally edible and possibly poisonous. What toxins and bacteria might be germinating in old and dirty whale fat?

The Fuegans, though they inspired the usually unflappable Darwin to write that the gap between "savage and civilized" humans could be greater than "between a wild and domesticated animal,"

[6]Charles Darwin, *The Voyage of the Beagle* (New York: Doubleday & Co., 1962), 214.

always carried fire with them, even in their canoes, for fear of being unable to start a new fire in their sodden homeland. These hold-overs from the Paleolithic knew what to do to make putrid blubber into food. They cut it into thin slices, roasted them over a fire, feasted—and survived.

2

AGRICULTURE

The man lay with his wife Eve, and she conceived and gave birth
to Cain. She said, "With the help of the Lord I have brought a
man into being." Afterwards she had another child, his brother
Abel. Abel was a shepherd and Cain a tiller of the soil.

—*Genesis*, 4: 1–2

Some forty or fifty thousand years ago, when the Upper Paleolithic surge was beginning, the world climate began to
warm; the continental glaciers, which covered all of Canada,
Scandinavia, and more, started to retreat. By 10,000 BP the relative
sizes and positions of the ice, the continents, and oceans were
roughly as they are today, and hunter-gatherers here and there were
teaming up with certain plants and animals to mutual advantage in
exploitation of sunlight. This was for humans chiefly a matter of
enormous multiplication in the kinds and quantity of photosynthetic products, direct and indirect, available as food, the fuel for
their prime mover: muscle.

The cause or causes of the thaw are a mystery. Perhaps there were
major alterations in prevailing ocean currents, which can have enormous effects on continental climates. If today the Gulf Stream
ceased to sweep northward out of the tropics, Europe would soon
have a frosty Labradorean climate. Perhaps there was an increase in
greenhouse gases like carbon dioxide, blocking the flow of heat into
space. Perhaps there was a change in the Earth's relationship with
the Sun, the most powerful of all influences on life on this planet.

The Earth's orbit is not a perfect circle, but an ellipse which changes shape over the millennia, increasing and decreasing the difference between the planet's nearest and farthest annual distances from its star. The tilt of the Earth's axis also slowly shifts, exposing different parts of the planet's surface to varying amounts of solar radiation. These and other variations in the Sun-Earth relationship manifest themselves in cycles of many thousands of years. Sometimes the cycles coincide and enforce each other and sometimes they cancel each other. Their effects can be increased or decreased by, to cite one example, the extent of ice and snow cover, which reflects radiation back into space, cooling the planet.

It is likely that there was a relationship between global warming, the origins circa 10,000 BP of agriculture, and the beginnings of the complicated phenomenon we entitle civilization. As the ice retreated, the curtain rose on the period called the Neolithic, during which some humans shifted from flaked to ground and polished stone tools, invented agriculture, weaving, pottery, and metallurgy, and began to live in permanent villages. These are the ingredients of what we vaguely refer to as "civilization." (The older textbooks usually dated the Neolithic as having begun about 10,000 BP and finished about 6,000 BP. As with the term "Upper Paleolithic," the dates vary by region and keep changing as archeologists make new discoveries.)

As the period opened, all our ancestors in total were only a few million hunter-gatherers dependent on wild plants and animals for the sun energy in food. A few millennia later, no more than a moment geologically and brief even by the standards of hominids' frenetic evolution, there were a lot more of them, the majority of whom were agriculturalists[1] dependent on domesticated plants and animals. Before Abraham of the book of Genesis was born (perhaps 4000 BP?), the archetype of Farmer Jones was already plodding along behind his ox and plow, munching, if he were a prosperous

---

[1]The terms "agriculturalist" and "agriculture" are often taken to refer to cultivation of plants and not of animals, but I use them to refer to both.

peasant, on some Neolithic precursor of a ham and cheese on whole wheat bread.

Men and women recruited certain plants and animals as means to mine for sun energy; or, to put it more precisely, humans converged in purpose and effect with these organisms. By 4000 BP almost every crop plant and animal essential to civilization today—wheat, rice, barley, potatoes, dogs, horses, cattle, sheep, and chickens—was domesticated. Since then we've added few of greater worldly importance than strawberries and reindeer.

By 4000 BP in several parts of the world dense populations of farmers supported city dwellers, hierarchies of specialists with skills like writing, and elites in government, war, religion, and manufacturing. With that, *Homo sapiens* nominated itself as the possible keystone species of the planet.

The word "domestication," as commonly spoken and written, seems a simple concept: it means taming of certain plants and animals, often with physical alterations beneficial to humans. For instance, we tamed wolves and made them into dogs. The word may seem inappropriate when applied to plants—would you say we domesticated potatoes?—but that is only a matter of verbal style. We seized species of wild potatoes, which probably had tiny tubers that tasted bad and may even have made eaters sick unless peeled, and bred them until they did our bidding, that is, we used sun energy to make big tubers that tasted good and made people fat and happy. But this definition and illustrations of domestication are overly flattering to us as human beings. They imply that we are in the driver's seat of our destiny, and we all know that that is far from completely true. In fact, the domestication of plants and animals isn't a unique human accomplishment like controlling fire.

We are inclined to think of cooperation between non-human species as a process of symbiosis, proceeding without intention; but there are some cases in which cooperation seems to be intentional. Several species of ants, for instance, engage in what we wouldn't hesitate to call domestication if the ants were us. They capture aphids, shelter and feed them, and lick a sort of honeydew off the tiny insects, milking them in effect. *Atta texana* ants chew leaves, roll them into balls, and "plant" them in underground fungi gardens. When mushrooms grow on the fungi, the ants eat the mushrooms. This procedure, which the Harvard entomologist E. O. Wilson calls "one of the major breakthroughs in animal evolution," may have been going on for as many as 50 million years.[2]

The aphids, fungi, and mushrooms get room and board; the formic ranchers and farmers get nourishment. So, you see, there are reasonable arguments that humans were not the pioneers in agriculture. Either ants can think creatively and plan ahead, or domestication can be the product of something other than reason.

We also must grant that the origin of agriculture was not particularly revolutionary, if by that we mean fast, as we use the word in our time. The most important achievements in domestication took several millennia, and the more that archeologists refine their tools and techniques, the further back the origins of agriculture tend to slide. If, for instance, you insist on actual plump squash seeds for evidence of that plant's first cultivation, then that happened about 5000 BP. But the evidence of phytoliths, microscopic bits of a glassy mineral, silica, from groundwater that accumulate in plant cells and take on their shapes, pushes the first domestication of rice from about 6000 BP all the way back to 11,000 BP.

And there is the eternal problem of definition. The Paiute of the Owen Valley of California used to dam streams to make wetlands for food plants like wild hyacinth, nut grass, and spike root, but they never sowed them. Were they farmers? Australian Aborigines dug up the tubers of wild yams to eat, replaced their tops in the soil so there would be more yams, and then went on their way. Were these foragers really farmers? The Sami (Laplanders) of northern Scandinavia and Russia lived too far from the sun—literally—to

[2]Quoted in Elio Schaechter, *In the Company of Mushrooms: A Biologist's Tale* (Cambridge, MA: Harvard University Press, 1977), 241–42.

tap it via domesticated plants. They depended on animals as their means to derive energy from the sun. As hunters, they padded along behind the wild caribou. They learned to direct the herds by throwing rocks, waving sticks, shouting, and best of all by getting the lead animal to go where they wanted, probably by tempting it with their own salty urine. The rest of the herd would follow—and then the Sami would follow them. Who was leading whom? When did the Sami hunters become herders? When did the wild caribou become tame reindeer?

Domestication at the beginning was not so much a matter of human intent as of human acceptance and exploitation of lucky accidents. Learning how to control fire had been revolutionary: the blazing branch did not of itself develop into a torch; somebody had to pick it up to make it a torch. Domestication, on the other hand, was in its beginnings evolutionary. Certain humans, animals, and plants evolved so as to become a team.

The change surely did not come because humans looked forward to the ultimate results. Saving a large portion of collected seed not to eat but to plant must have initially struck a lot of them as counterintuitive. Living with and taking care of big animals like cattle and pigs for future benefit, rather than killing them on sight for immediate consumption, may have struck them as particularly strange.

The hunter-gatherer way of exploiting sunlight had been successful. It had carried us from our continent of origin, Africa, all the way across Eurasia to Australia and the Americas. There weren't very many of us circa 10,000–15,000 BP, but more than ever before, and, if we accept the testimony of bones, we were all in all healthy; and, if we accept the testimony of artifacts of technology and art, a creative and, possibly, a happy lot. Why did we drop foraging for new ways to tap the sun?

The hunter-gatherers of our day are our equivalents of the people of the Upper Paleolithic. The normal diet of a Bushman of the Kalahari Desert in good times is an enviable 2,000-plus calories daily. In bad times, it is a good deal less, but Bushmen seldom starve to death. Long ago they identified about seventy-five plants and sixty animals that are edible. One Kalahari hunter-gatherer, when asked why he and his people didn't become farmers, answered, "Why should we, when there are so many mongongo nuts in the world?"[3]

There seemed no immediate gain to dropping the spear and picking up the hoe, so why did the first farmers do so? The usual explanation credits the origins of agriculture to global warming, which altered regional climate and rendered traditional ways of food procurement obsolete; but that may have been no more than what logicians call a necessary condition, not a sufficient cause. There had been such periods before, during the Paleolithic, which had *not* triggered farming.

Perhaps humans themselves forced the transition. The people of the Upper Paleolithic were superb field ethologists, with a new and improved kit of weapons. At about the same time as agriculture's first stirrings, 11,000 or 12,000 BP, most of the big herbivores of the Americas—mammoths, horses, woolly rhinoceroses, and ground sloths—went extinct, as did similar animals in much of the Eastern Hemisphere. Maybe the hunters eliminated all of these dumb lumbering meat loaves. (They didn't actually have to kill all of them, but only to kill more than were born for a century or two. Big animals—known in scientific jargon as megafauna—have very low birth rates but 100 percent death rates, just like mice and mites.) Then there weren't enough megafauna to eat, so one sunny afternoon our ancestors invented agriculture.

Possibly, but the big animals survived quite nicely in Africa (elephants and lions, for instance) and in southern Asia (elephants and tigers). Furthermore, there were regions where these giants disappeared but agriculture did not follow: Australia would be the clear-

[3]Stephen Budiansky, *The Covenant of the Wild: Why Animals Chose Domestication* (Leesburg, VA: Terrapin Press, 1995), 115.

est example. And there is the chronic problem of scarce evidence. Archeologists have identified only a few ancient sites where hunters killed and butchered mammoths and such. That does not prove the megafauna hypothesis wrong (to quote an old aphorism, lack of evidence is not necessarily evidence of lack), but it certainly doesn't serve to confirm it.

Perhaps human population pressure propelled the shift to agriculture. Hunter-gatherers, like lions and tigers, need enormous spreads of territories to exploit, something like ten square miles per person, because the plants and animals they depend on to stay alive rarely occur in dense concentrations. For hunter-gatherer bands, the sight of other bands can be a portent of famine. *Homo sapiens* had dealt with that problem previously by expanding geographically from Africa all the way to Australia and then to America and south to Cape Horn. It may be that by about ten thousand years ago humans had filled up the world with as many people as could be supported with hunter-gatherer technologies. That obliged them to do something new and clever about getting enough to eat—to invent agriculture, for instance. Farmers tap sun power so efficiently that they need only about a tenth of a square mile to support a single human.

That explanation is appealing in that it addresses the scattered locations of the lands in which agriculture originated. As with climate change, humans could have become conscious of population pressure in many different parts of the world in the same millennia. But (and you know what comes next) there is very little physical proof for this. Archeologists have not found circa 10,000 or 12,000 BP, not even in the Middle East where agriculture may have first blossomed, the crowding together of living sites that one might look for as evidence of population pressure, and there is little evidence in human bones of a decline in the quality and quantity of nutrition, often a side effect of population pressure.

Perhaps none of the hypotheses above is valid. Perhaps all are valid. Nature is not a laboratory scientist obliged to test for one factor at a time. The climate change that ended the Ice Age circa 10,000 BP may have rendered people's traditional practices obsolete, and the decline of the megafauna mentioned above surely made life harder for hunters. There certainly were more humans in total than ever before because members of that species had migrated to almost all the lands where they would ever live. We know that techniques for exploiting new foods, such as grinding hard seeds, developed and spread. We know as an archeological certainty that humans in many regions intensified their exploitation of plants and smaller animals, learning not only how to harvest both, but how to encourage their propagation and availability for future meals.

Let's add to the roar of debate one more hypothesis, one for which there is physical evidence, as well as support from the plant and animal sciences: humans were not leaders but only participants in the drama of the first domestications. Certain plants and animals evolved to suit humans—and so did humans to suit plants and animals. Domestication was not a matter of humans dictating to subordinate organisms, but of people and plants and animals evolving in physical structure and composition, life schedules, and so on in reaction to each other. Evolution proceeded in accordance to the principle of the survival and reproduction of the fittest *team*!

There seems to have been in some areas a slowing of the geographical velocity—the seasonal, even daily, geographical jiggle—of humans. Hominids and humans had always been transients who provided little in the way of a fixed *mise en scène* for interaction of humans and other organisms: plants, animals, and, yes, germs. Humans had been like sand in moving water, sweeping along with the flow. Now their pace slowed as their technologies and social skills enabled them to forage more efficiently. The sand settled and became shoals and islets—villages—and flotsam began to accumulate.

The classic hunter-gatherers move *through* environments, altering them, yes, but seldom drastically because their presence in any given location is temporary. Villagers, in contrast, live *in* their surroundings and change them, stimulating changes in resident organisms. Villagers collect wood for fires, tools, and structures. They create a sunny hole in the forest by burning off surrounding trees and bushes so they can spot predators, animal and human. They kill or scare off the local big game and root up many of the edible plants. They expel some animals, like ferrets, and unintentionally encourage the kind that happily feed on offal and garbage in general: mice, for instance. Humans have always had a talent for creating dumps—middens is the formal name—which attract all sorts of minions. They and the humans joined in unconscious alliance to start domestication. Let us consider a few illustrations of how these organisms teamed up, beginning with an animal of enormous significance for human beings, the dog.

The domesticated organism with which we humans are most anciently and intimately associated, excepting the bacteria who live in our gut, is *Canis familiaris*. Recent research in DNA obtained from dogs all over the world informs us that all our dogs are descendants of wolves, probably East Asian wolves. The DNA further suggests that this domestication happened as long ago as 40,000 and no more recently than 15,000 BP. Since there is no physical evidence for canine domestication, such as the appearance in association with humans of a "wolf" with a shortened muzzle and crowded teeth, later than roughly 10,000 BP, 15,000 BP seems a good guess.

The extraordinary ability of dogs and humans to communicate suggests that they have been close for at least that long. Dogs are better than chimps, our closest cousins, at interpreting our voice and hand signals, and they possess that ability at ages too young for it to have been the result of training, intentional or unintentional. It also seems likely that we have evolved to send messages to dogs and to understand the messages that come back. According to current research, "aspects of the social-cognitive abilities" of dogs and humans have converged.[4]

[4]Brian Hare et al., "The Domestication of Social Cognition in Dogs," *Science*, 298 (Nov. 22, 2002), 1636.

Consider who dogs and people were in 15,000 BP. The "dog" was a wolf, a big carnivore. The human was a hunter and a big omnivore who dined on meat as often, probably, as possible. They were rivals for the same kind of food and must have been as willing to kill and eat each other as are sharks and fishermen today. They weren't, it would seem, a likely pairing.

The stock explanation for the domestication of the dog is that humans initiated the change by adopting wolf puppies, bringing them up as pets, and breeding them for gentleness and obedience. This is a practice that doesn't always or even usually work with most animals. Humans have tried the technique on moose, raccoons, bears, gazelles, ibex, hyenas, antelopes, and American buffalo with no success. As these animals grow up, they become unpredictable, panicky, dangerous, fiercely territorial, and won't or can't breed in captivity. Their "social-cognitive abilities" don't, probably cannot, converge with humanity's. How is it that dogs and humans got along together so well?

Let us return to our village of the late Upper Paleolithic, our islet on which drifters can land and stay. Humans make garbage. Wolves like some kinds of garbage and they especially like some of the kinds of animals that like garbage. They drift in closer and stay longer and become quasi-residents (like mice, only bigger). People tolerate them because they can't make them go away and stay away and because wolves make a lot of fuss when strangers, animal or human, come near. Wolves turn out to have behaviors useful for humans, just as the garbage-making latter do for the former. Wolves have keener senses than human hunters and are our superiors tactically, but humans are smarter strategically. Wolves in pursuit of meat are better hunters *right now*; humans know better where to go and when to go there for a generous acquisition of protein *next season*. Both kinds of animals live not as isolated individuals, like orangutans or cougars, but in packs. Both tend to accept and to operate well in hierarchies. Humans will follow their hunting dogs. Dogs will accept humans as pack leaders. Sometime not less than 15,000 BP they joined up with each other, and each has been better fed, safer, and the other's best friend ever since. This bonding is the best example of the journalist Stephen Budiansky's proposition that "domesticated animals chose us as much as we chose them."[5]

Let's look at another example of mutual domestication, horses and people. They were not necessarily rivals for the same kinds of food and did not therefore begin by hanging around each other, so their domestication came late. Humans and horses started as enemies, with the former as the hunter and the latter as the food. Horse bones, chopped and scored with knife marks, are plentiful in middens of the early Neolithic. Indeed, horses may have been originally domesticated as food.

Horses are big and, though herbivores, dangerous, especially the stallions, but otherwise the species was a good candidate for domestication. Horses aren't habitually panicky. They aren't fiercely territorial; they couldn't be because they are always on the move. They, too, function in packs (or, to use as word more suitable for this species, herds) with a single stallion as *el jefe supremo* and one mare as leader of the harem. When they migrate, the stallion exercises control from the rear of the band and the highest ranking mare leads. They can be accustomed to accept a human as chief.

Eventually these powerful and often amiable animals attracted the attention of humans tired of carrying all their burdens themselves, tired indeed of carrying themselves. About 6000 BP lazy humans living on the Ukrainian steppe transformed these meat sources into pack animals, wagon and chariot pullers, and soon climbed onboard the big quadrupeds themselves. In return, humans guided their horses to sunny pastures, sometimes brought them straw and grains when the grasses failed and provided shelter in bad weather. Humans may even have saved horses from extinction.

Horses were very nearly included among the megafauna that disappeared in the late Pleistocene. They did disappear entirely from

[5]Budiansky, *The Covenant of the Wild,* 24.

the Americas where they had first evolved, and from much of the expanse of the Old World where they had lived in large numbers when mammoths walked the earth. It is tempting to think that human hunters were the cause of their decline because horses disappeared from North America at approximately the same time that expert Upper Paleolithic hunters appeared there. Be that as it may, the area where wild horses lived shrank to the semi-arid lands of central Eurasia.

Humans intervened to reverse this retreat and decline. Two millennia after domestication, horses were back in Western Europe—as livestock. Five or so millennia after that they were back in America, some of them still domesticated and more of them gone wild again, overflowing the ecological niche that native horses had vacated. The North American word "mustang" comes from the Spanish *mesteño*, referring to a horse or steer that is ownerless, wild.

Livestock, especially horses, were a massive acquisition of immediately available speed and power for humans, and we were impressed. Even now, thousands of years later, we find a dictator or a movie star mounted on a horse particularly heart-throbbing. But the dull and dry fact is that most of us for at least the last ten thousand years have derived more of our nourishment (remember, that means sun energy) from plants than animals. Let's look at domestication in the former category, most particularly at grasses, confining our attention to the calorically generous staples alone. (Provide humans with the minimum requirements for calories and they will usually manage to scrape together enough of the other essential nutriments to keep going and reproduce.)

Grasses have been staples for many, perhaps most, people since the Neolithic: wheat, barley, rye, rice, oats, millet, sorghum. (Sugar is another grass of huge historical influence, but not a staple for hundreds of millions of us until the coming of tea, coffee, and of course Coca-Cola.) Grasses are opportunistic. They make hay while the sun shines, and seeds, too. They move in fast to occupy bare ground, such as the stripped countryside around the early villages.

They complete their cycle—produce food—in less than a year. Their seeds are nourishing and tolerate storage well. Grasses tend to grow in solid stands, simplifying the task of creatures feeding on their seeds.

Take wheat, for example, a plant of enormous importance in human diet. Wild wheat was an important food for Middle Easterners far back in the Upper Paleolithic, but there was a perennial problem with collecting its seeds. The labor of picking them up one by one out of the dirt and humus canceled out much of their value as energy sources. The kernels of wild wheat are fixed to the stem with brittle connections. When disturbed by wind or, in this case, human hand, the head with the seeds "shatters" and scatters the seeds to the ground away from the plant, maximizing chances for germination and the propagation of the next wheat generation. Only the occasional odd mutant doesn't shatter; these seldom succeed in reproducing because all the seeds of a stalk fall to earth in one spot and try to take root there, crowding out each other. They have little chance to outcompete normal wheat plants.

Human gatherers unintentionally collected more kernels of the non-shattering variants because these stayed on their stalks. A sweep with a flint-toothed sickle sprayed the normal seeds into the dirt, while the non-shatterers' remained on their stalks neatly laid out to be picked up. This had little long-run significance for the foragers who were passing through, but for villagers it altered the course of their lives. They brought the non-shatterer seeds back home, where they milled, soaked, and cooked them. They and their relatives ate most of the seeds and infastidiously spilled the rest or threw them into the middens. In the next wet season more non-shattering wheat sprang up in the vicinity of the villages. The women who were the gatherers favored these fields of wheat because they were close by. Next season yet more non-shatterers sprouted. At some point human intent replaced accident and wheat farming ensued.

The law of survival of the fittest (in this chapter that means the organisms that thrive in our vicinity) altered strains of cultivated

Workers harvesting with sickles; detail from the tomb of the scribe of the fields and estate inspector under Pharaoh Thutmose IV, Thebes, Egypt, sixteenth to fourteenth century BCE.

wheat further. The plant no longer, so to speak, hedged its bets, as its wild ancestors did. Take, for example, the matter of germination timing. The thousands—millions—of cultivated wheat plants in a given field all produced seed simultaneously; none of them held back to sprout another year for fear of drought or early frost or what-have-you. Cultivated wheat marched in lockstep for the convenience of the reaper, who returned the favor by broadcasting the seeds of just this variant next season. One expert has said, plain and simple, "Wheat and people co-evolved in ways that left neither much ability to manage without the other."[6]

The Neolithic changed humans just as it changed livestock and

[6]John H. Perkins, *Geopolitics and the Green Revolution: Wheat, Genes, and the Cold War* (New York: Oxford University Press, 1997), 19.

crop plants. Humans may have undergone many genetic adaptations that we do not recognize because they are common to all or most of us. We are limited to considering those distinctive to minorities of our species. Consider, for example, the matter of the toleration and consumption of milk by adults. Humans, like mammals in general, live on milk for considerable periods following birth. Then the majority of us are weaned and between about two to five years of age lose the ability to digest lactose, milk sugar. Most adults who drink milk suffer gas, bloat, and nausea, and therefore sensibly forgo milk or consume it safely as cheese, yogurt, and such.

Lactose-intolerance is the rule for Asians, Africans, southern Europeans, American Indians, and Pacific peoples. The exceptions to the rule are restricted to a minority, including many Middle Easterners, possibly the Tutsi of eastern Africa, and almost all Northern Europeans and their descendants. There is a clear correlation between where these lactose-tolerant peoples live today, or lived until recent migrations, and where humans have been dairying for millennia. The Darwinistic scenario for Northern Europeans has their ancestors suffering rickets because their cloudy climate cut off sunlight, a rich source of a necessity, vitamin D. Survival of the fittest meant survival of those few who could derive vitamin D from milk as adults. *Mirabile dictu*, the propagation of a subspecies of pale people swilling milk, pouring it on their cereals, and licking ice cream cones.

Agriculture changed humans in ways parallel to their livestock, if not for the same reasons. The latter, with the exception of unintimidating species like the chicken, shrank in size as their keepers killed off the big and ferocious and bred the survivors with each other. Humans also diminished: skeletons of adult inhabitants of Greece and Turkey of the late Upper Paleolithic average five foot nine inches for males and five foot five inches for women. The heights of men and women living in those regions after the shift to agriculture dropped six inches and five inches, respectively, and tooth decay increased for both sexes. Peoples all over the world, as they shifted from hunter-gathering to farming, suffered similarly.

Hunter-gatherer diets had been varied and richly nourishing—when available. The diets of farmers, of whom there were many more than pastoralists, consisted of cereal grains and roots, with meat or fish only now and then. These diets were much less nourishing, and brought on dental decay, anemia, and other maladies of malnutrition. Added to these disadvantages were the infectious diseases—smallpox, measles, dysentery, cholera—that sedentary life, dense population concentrations, and constant sharing of parasites and germs with dogs, horses, cattle, and pigs encouraged. We can say, all in all, that civilization was founded by runts with cavities.

Yet agriculture, for all its grim characteristics, spread swiftly across the Old and New Worlds. A more efficient exploitation of solar energy than hunting and gathering, it enabled more people to live on less land. Absolute starvation might be more common among the farmers than foragers, but in most years more babies were born to the farmers because in most years agriculture provided more food. In addition, farming was a comparatively sedentary activity and mothers did not burn up all their carbohydrates and fats lugging offspring from place to place. As a result they were more fertile more often and, as well, resorted less often to infanticide and other techniques to limit the number of mouths to feed.

The agriculturalists' dense populations sometimes advanced into the hinterlands aggressively and sometimes spread passively like piles of sand sliding downhill. The hunter-gatherers who lived adjacent to the lands of the agriculturalists either adopted agriculture or were absorbed or erased. When Moses led the Israelites out of Egypt, the majority of our species were surviving by plugging into the sun via domesticated plants and animals.

A big population such as that formed by agriculturist groups was a prerequisite for a complex society and what we call civilization. Dense concentrations of people did not automatically or swiftly produce new prime movers; but simply by existing they increased the quantity of muscle, the existing prime mover, available to do work. Ziggurats and pyramids were built by multitudes living on sunlight tapped by domesticated organisms.

CODA: DOMESTICATION IS NOT AUTOMATIC, AN ALASKAN EXAMPLE

Today, the advantages of agriculture in tapping the sun seem to its beneficiaries so obvious that we slip into the assumption that its spread must have been nearly automatic. Our oversimplification is the effect of post-Neolithic self-esteem and the common sin of foreshortening the past, of flipping through the distant millennia like pages in a book because the evidence that has survived from each thousand years is so slight. But domestication was often discontinuous, didn't always work, and the spread of its organisms and practices has often been halting. For an example of that, consider the attempt to establish the most northerly of the livestock of the Neolithic, the reindeer, in Alaska.

Reindeer are caribou, a kind of deer with concave hoofs that are good for walking on and digging in snow. (*Caribou* means "shoveler" in the language of the Micmac of eastern Canada.) This big quadruped ranges through the tundra and boreal forests of Eurasia and North America. Its diet is varied in the warmer parts of the year, but in the winter it depends mostly on "reindeer moss," which isn't a moss at all but a kind of lichen that derives its nourishment from air and light and lives on the surfaces of rocks and trees. These animals have provided humankind with nourishment, skins for clothing, and bones and antlers for all sorts of tools since far back in the Upper Paleolithic.

Since there are no obvious differences between the bones of wild caribou and domesticated caribou (reindeer), we have little idea of the date when some caribou became tame. We know that it happened in the Old World long after the domestication of cattle, horses, and the other standard citizens of our barnyards. We know that reindeer have provided Eurasians of the far north with meat and milk, and with material for clothing and tools. They have pulled sleds and wagons and carried loads and even people for thousands of years, but we don't know how many thousands. They have been even more important to boreal Eurasians than domesticated animals have been to people in warmer climes because the northerners cannot count on the productivity of farming.

The last wave of the Inuit (Eskimos) to migrate from Siberia into North America did so prior to the domestication of the reindeer. The concept of domestication crossed the Bering Straits with dogs later, but if so, that made no difference because American caribou, though a herd animal, was and is distinctively resistant to domestication. And so matters stood *vis-à-vis* American Inuit and domesticated animals for scores and scores of human generations. Russian annexation of Alaska in the eighteenth and early nineteenth centuries brought it post-Neolithic weaponry, the fur trade, and the diseases of civilization, but not reindeer. Then, in 1835, American whalers (we'll get back to them again in the chapter on petroleum) entered the waters between the north of Alaska and Siberia, and within a few decades so reduced the whale population there that the whalers had to turn to hunting other sea mammals, walruses in particular, to maintain profits. They killed hundreds of thousands, leaving the coastal Inuit who depended on hunting them to go hungry.

In 1867, the United States bought Alaska from Russia. Among the few Americans who noticed that the Inuit had become an American responsibility were the missionaries, who went north to save their souls and to make white folks of them as nearly as that might prove possible. In 1885, the Reverend Sheldon Jackson, who had already spent a dozen years as a missionary among the Native Americans from Montana to New Mexico, became superintendent of public instruction for Alaska. He wanted to convert the Inuit and teach them to read, write, and calculate, but also knew they needed food and purpose. He campaigned for funds, private and federal, to import reindeer from Siberia, sure that herds of domesticated caribou would do for Alaska what cattle, sheep, and horses had done for the western states of the "Lower Forty-eight."

Jackson proclaimed that northern and central Alaska could support over 9 million reindeer. They would transform worthless wilderness into valuable real estate. They would "introduce large, permanent, and wealth-producing industries" where none had existed before. Most important of all, they would "take a barbarian people on the verge of starvation and lift them to a comfortable self-support and civilization. . . ."[7] The indigenes of Siberia, closely related to the Inuit in physique and culture, had been successful reindeer ranchers for a very long time, and the Alaskan natives would surely follow their example.

Between 1891 and 1902, a total of 1,280 reindeer were shipped from Siberia to Alaska, initiating a biological roller-coaster ride. Reindeer herds, if healthy and well fed, can increase by one third annually. In 1932, there were 600,000 reindeer in Alaska grazing from the north coast on the Arctic Ocean south to Kodiak Island and the Aleutians, all descendants of the initial few. Then their population plunged. In 1940, there were 200,000. Ten years later, they were gone from the North Slope and the islands and limited to a few patches of western Alaska. There were only 25,000 left. Sheldon Jackson's dream of easeful domestication of reindeer and Inuit failed spectacularly. Explosive ascent and then a crashing decline—how, why? Population explosions are normally prevented by competition and predation. Perhaps the wild caribou herds of Alaska had been thinned out by hunters using imported rifles. Perhaps their predators, the wolves, had fallen in numbers along with the big deer. We don't know the answer, but it is obvious that the reindeer moved into a wide-open ecological niche. The Inuit, who were supposed to

[7]Quoted in Norman Chance, "The 'Lost' Reindeer of Arctic Alaska," http://arcticcircle.uconn.edu/NatResources/reindeer.html. (Viewed May 28, 2004.)

spring for the opportunity to acquire their own reindeer, did not take to herding enthusiastically. They had not participated in the long negotiation of domestication with the reindeer. To put it another way, the reindeer were domesticated to serve humans, but the Alaskan natives were not domesticated to serve the reindeer. They were hunters, not herdsmen, and found their ancient ways more satisfying than the new ones abruptly offered them. The few who might have been inclined to enter into the capitalist economy took little interest in reindeer because the market for its meat was tiny.

Most important of all causes for the reindeer population crash was their extravagant success in reproduction. Some 600,000 reindeer might find enough to eat in Alaska in summer, but not in winter. In warmer climes, overgrazed ground cover may recover in months, even weeks; but not in the far north, where winter temperatures are abyssal and the sun bobbles along the horizon for a few hours daily or sinks beneath it and stays there. Recovery of ground cover there may take as long as thirty years. The hordes of imported deer stripped the reindeer moss from the tundra and forest and then starved.

Reindeer did not completely disappear from Alaska. Today, there are herds in the thousands in the Seward Peninsula and elsewhere, but they play a minor role in the Inuit and Alaskan economies. The Reverend Jackson's prophecy of "large, permanent, and wealth-producing industries" has not come to pass. Domestication is not preordained. Sometimes it works and sometimes it doesn't.

3

THE COLUMBIAN EXCHANGE

*All the trees were as different from ours as day from night, and
so were the fruits, the herbage, the rocks, and all things.*
——*Christopher Columbus, mariner (1492)*[1]

When Jesus and Mohammed walked the Earth in the
first centuries CE (Common Era), the towering
majority of humans were members of societies that
exploited the Sun via domesticated plants and animals. But as yet
no single division of humanity tapped the full advantages of agriculture because access to the total sum of Neolithic innovation was
limited by climatic contrasts and geographic distances, not to men-
tion mountain ranges, deserts, jungles, and especially oceans.

The *Homo sapiens* species had "invented" agriculture in at least
nine different regions: Southwest Asia, Southeast Asia, North China,
South China, and sub-Saharan Africa in the Old World; eastern
North America, central Mexico, the South American highlands,
and the South American lowlands in the New World. At first, each
of these centers had different sets of domesticates, plant and animal.
For instance, central Mexico domesticated maize, squash, beans,
and very little if anything in the way of livestock; South China
domesticated rice, pigs, chickens, and water buffalo.

The full exploitation of sun energy through agriculture, during
the era when muscle was humanity's foremost and nearly only prime

---

[1]Alfred W. Crosby, *The Columbian Exchange: Biological and Cultural
Consequences of 1492* (Westport, CT: Praeger, 2003), 4.

mover, waited on sharing among the nine centers of agricultural innovation. For thousands of years the sharing of techniques and crops went on slowly and entirely within the two continental masses of the Old and New Worlds. For instance, rice and sugar cultivation migrated from Southeast Asia to India to the Middle East and then to Europe at something well less than a walking pace. Maize in Mexico dates as far back as 9000 BP, but wasn't grown in the eastern United States until about seven thousand years later. Sorghum, a grain, was cultivated 5000–6,000 BP in Africa, got to India no sooner than two thousand years ago, and from there on to China, perhaps via the Mongol conquerors, circa the thirteenth century.

Sharing within both the Old and New Worlds was slow, but was well underway in the second millennium CE, and millions of people were dependent on domesticated plants and animals that had first been recruited for human service far, far away. In contrast, no cultivar or domesticated animal had crossed the Atlantic or Pacific and become established on the far side. The South American lowlands' sweet potato somehow spread from there through the Polynesian Pacific, but not all the way across to Asia; and Asian rice, an excellent crop for tropical Pacific islands, one would think, somehow did not spread through Polynesia at all.

To consider this long-lived insularity of the Old and New Worlds, we return to the subject of global warming circa 10,000 BP. The sunlight streaming down warmed the globe and the continental glaciers melted. Their water flowed into the oceans and sea levels rose, drowning the land connections between the British Isles and continental Europe, New Guinea and Australia, and, most portentously, between Northeast Asia and Alaska, that is, between the Old World and the New.

Humans had yet to build seacraft capable of crossing oceans dependably and repeatedly—even the ancestors of the Polynesians were novice seafarers in 10,000 BP—but people had already traveled from Siberia to Alaska on foot over the Bering land bridge (a "bridge" as broad as Texas) and into the Americas. They brought

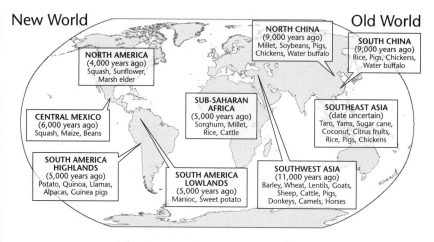

The Separate Inventions of Agriculture

with them the cultures of the Upper Paleolithic. These included all sorts of stone crafts, cooking techniques, such devices as the atlatl (throwing stick), and the dog (but not, as we have already seen, the reindeer). Then the sea level rose, Beringia—as the land bridge has been posthumously entitled—disappeared, and these first Americans and their dogs were left alone before the full blooming of agriculture anywhere.

Fitful contacts between the Eastern and Western Hemisphere peoples seem to have continued. Bow and arrow technology passed from the former to the latter after the inundation, but little else. But the connection of Siberian and Alaskan chilly tundra could never have served effectively as a corridor for the exchange of plants and animals between the warmer latitudes of the two hemispheres, where the shift to agriculture was beginning. Between the flooding of Beringia and the ocean ventures of Christopher Columbus, Old and New World food procurers lived and evolved independently behind broad moats, and would continue to do so until somebody began crossing the oceans repeatedly and dependably.

To cross oceans, the prime mover, muscle power, won't suffice.

Animals cannot be trained to row efficiently and for long periods. Humans can be, but that approach won't work, either. Rowing on lakes or the Aegean is fine, but if you enlist (or enslave) enough men to row a ship across an ocean, you'll have to take on board a lot of food, which will necessitate a bigger ship, which will require more rowers, who will also have to eat to function, and so on and on. Rowing across oceans is not a practical means of travel. Fortunately, the Sun provides another means.

The Sun wherever it shines heats bodies of air differentially. These bodies expend that sun energy scurrying and blustering about as winds, swerving in obedience to the Earth's rotation. Wind is a form of sun energy available on land and sea. In the centuries between Mohammed's death and Columbus's birth, mariners and shipbuilders of the China Sea, Indian Ocean, Mediterranean, and Europe's Atlantic coasts learned how to build large and truly seaworthy vessels, and to sail them, tacking against the wind if necessary, across the major oceans, utilizing the compass to find their way. The Chinese jumped ahead at first, dispatching enormous fleets as far as East Africa, but then reverted to traditional and continental concerns. The Europeans, with the Portuguese and Spaniards in the vanguard, took the lead. They were as good as any continental peoples in their seamanship and navigational skills, and more adventurous.

The Portuguese and Spaniards learned how to tap the sun energy of the winds not only locally, but oceanically. They discovered the great clockwise systems of air flow that dominate the surfaces of the Atlantic and Pacific: easterly trade winds in the tropics and prevailing westerlies in the temperate zones. They utilized this knowledge to ride the trade winds from Iberia to America, from Acapulco to Manila, and then back again via the westerlies. While doing so, they carried the domesticated crops and livestock of the Old to the New World, and vice versa. (They also brought diseases like smallpox and malaria to America for the first time, killing millions. The Native Americans slyly countered with tobacco.)

Ships depicted by Theodor de Bry in *India Orientalis*, 1605.

Agriculturalists on opposite sides of the Atlantic and Pacific worked with different sets of plants and animals. All the animals mentioned thus far—dogs, cattle, horses, sheep—were from the Old World. The people who would become Native Americans almost certainly arrived in the New World with dogs, so they were familiar with the concept of partnership with animals. It is not clear why they didn't domesticate more animals, or at least more useful animals than they did. They certainly adapted to horses and cattle swiftly after 1492. Perhaps the megafaunal extinctions of the Upper Paleolithic mentioned in chapter 3, which were more thorough in the New World than the Old, eliminated the likeliest local candidates for domestication such as the American horse. Native Americans established close alliances with the guinea pig, some fowl, but no large animals except the llama, which only lived in Andean South America. They had no ally they could ride or hitch to a plow, and none that they could depend on for milk. But Native Americans excelled as farmers. Of the major crop plants that serve today as staples of our species, a third to a half are of New World origin. This is remarkable considering that the Old World includes most of the planet's dry land and, when agriculture was first evolving, surely was home to as great a proportion of human beings.

Before we pursue that point further, let us appraise the Old World contributions to the New World agriculture. In the first few post-Columbian centuries, the imported cultivated plants effected great changes in those parts of America for which Old World experience had pre-adapted them. There were many localized successes—

African slaves planting sugar cane under European supervision.

olives, wine grapes, peaches—but let us focus on the continental or nearly continental successes. Sugar cane, which originated in New Guinea, did wonderfully in the wet American tropics from the West Indies to the hump of Brazil. For the first time an intense sweet, which humans, like their apish ancestors, were genetically pre-adapted to like, was available in quantity. Europeans developed a sweet tooth and millions of Africans were enslaved and shipped to the New World to work the cane fields to satisfy that taste.

Let us consider wheat as another example. In the nineteenth century, the influence of Euro-American capitalism flowed inland and penetrated the North American plains and South American pampas. Its Myrmidons sowed a southwest Asian grass there—wheat—harvested the grain, and began shipping it out to the world to feed the urbanized labor forces of the industrial revolution everywhere. Wheat is still the biggest food item of transoceanic trade.

The impact of Old World livestock on the New World was even greater than that of its cultivars in the short run and possibly in the long run, as well. Celebrities of the Old World Neolithic, mobile recipients of the sun's largess—horses, cattle, pigs, sheep, goats, and chickens—transmogrified the American biosphere and were running loose in the millions within two centuries of the Columbian voyages.

Pigs led the way because most of the first European settlements were in the wet, well-wooded lands where they thrived, multiplied, and went wild. Soon after the founding of the Spanish colony in Española, their numbers were *infinitos.* "The swine doe like very well heere," a visitor to Brazil wrote in 1601, "and they begin to have great numbers, and heere is the best flesh of all." In early Virginia, they "did swarm like Vermaine upon the Earth."[2]

Cattle are not as well suited to the damp tropics as pigs, but they multiplied explosively after a while. A 1518 report to the king of Spain stated that thirty or forty stray cattle in the West Indies would

[2]Ibid., 79.

increase to hundreds in as little as three or four years. When cattle reached the grasslands of northern Mexico and Argentina, their numbers soared into the millions. One careful observer estimated the number of wild cattle in the pampa alone circa 1700 at 48 million.

Horses, a finicky species compared to pigs and cattle, lagged behind, but not far behind. By the end of the sixteenth century the plains of Buenos Aires were "covered with escaped mares and horses in such number that when they go anywhere they look like woods from a distance."[3] By the mid-eighteenth century they had reached Canada from Mexico, via Native American rustling, trading, and their own independent drifting.

Livestock provided the inhabitants of America with food in large quantities as well as new sources of fiber and leather, and of strength. The enormous pyramids, buildings, and monuments of Mexico before Spanish conquest had all been raised by human muscle alone; now there was oxen muscle to help. No inhabitant of America had ever traveled overland faster than he or she could run. Now they mounted horses and galloped like the wind.

The impact of New World livestock on the Old World, on the other hand, has been nil. Guinea pigs have proved useful in medical research and turkeys have graced a few tables, but nothing more significant than that. In contrast, the influence of New World crops there has been enormous. Let us turn to the sunshine staples, the calorie generators on which hundreds of millions in the Eastern Hemisphere have depended for at least two centuries. The most important are white potatoes and maize.

White potatoes, often called Irish potatoes, came from northern Chile and Peru, where there are many species, wild and domesticated, obviously related to this cultivated tuber. Potatoes provide a variety of nutrients—vitamin C and even some protein—and are closer to being sufficient to support life all by themselves than any

[3]Ibid., 85.

other major crop plant. But historically they have been chiefly important for their calories. In cool, damp climates they produced more calories per acre than any Old World crop and became a staple for the lower classes. In 1776, no less a seer than Adam Smith paid the potato its due in a backhanded compliment to the Celts:

> The chairmen, porters, and coal-heavers in London, and those Unfortunate women who live by prostitution, the strongest men and the most beautiful women perhaps in the British dominions, are said to be, the greater part of them, from the lowest rank of people in Ireland, who are generally fed with this root. No food can afford a more decisive proof of its nourishing quality, or of being peculiarly suitable to the health of the human constitution.[4]

The cultivation of "this root" (it is a tuber, not a root) requires no more complicated or expensive tools than a spade. If you don't have a spade, a digging stick will do. Potatoes have the advantage of keeping their edible parts underground, safe from feral livestock and all but the most determined thieves.

The origin of maize seven to ten thousand years ago is one of archeo-botany's deepest mysteries. Cultivated wheat is like wild wheat in appearance, but maize seems unlike any plant growing wild in its homeland, Mesoamerica (central Mexico and regions to the southeast). Perhaps it evolved so swiftly and radically under human influence that it doesn't resemble its ancestor at all. Perhaps its ancestor is extinct. Perhaps maize turned Oedipal, crossed with its parent, and simply absorbed it genetically.

The first cultivated maize of which we have examples bore cobs about an inch long and as thick as a pencil, with four to eight rows of kernels. The total food value of an average cob may have been less

---

[4]Alfred W. Crosby, *Germs, Seeds, and Animals: Studies in Ecological History* (Armonk, NY: M. E. Sharpe, 1994), 157.

than one kernel of maize today. Fortunately, maize is one of the most genetically flexible of all crop species and in due time Native American plant breeders had it producing big cobs with plentiful kernels. They cultivated it from Canada to Argentina. Today, it turns out more calories per acre than any grain but rice.

Maize was and is easy to cultivate, requiring in planting and cultivation no special tools or greater skill and strength than a ten-year-old might possess. Harvesting requires no tools but hands. At maturity the ears, encased in tight wrappings of leaves, can be left on the stalks until it is convenient to pull them off. Maize kernels, unlike potatoes, can be safely stored for years. The exiled American loyalist Count Rumford (Benjamin Thompson), reminiscing about maize, praised it as "beyond comparison the most nourishing, cheapest, and most wholesome [food] that can be procured for feeding the poor."[5] (Peasants, often dependent on foods produced on infertile lands of marginal utility, e.g., alongside paths between rice paddies, were usually the earliest to commit their fates to American crops.)

In return for nourishing humans, maize, like the horse, won survival. It is as dependent on farmers as are the tapeworms in their bellies. Maize kernels (seeds) left to their own pursuits rarely escape their sheath of leaves, and if they do, cannot scatter. Maize is an even more extreme example of the interdependence of plant and human than wheat.

The most important thing about the different crop plants of the two independent Neolithics is just that: they were simply *different*—with different advantages and disadvantages, different ways of dealing with challenges to their growth and reproduction. Imagine the pre-Columbian populations of the Old and New Worlds as two individuals playing poker with nature, "betting against the house." Each has five cards and wins once in a while. After 1492, they shared their cards. With ten cards collectively, they were able to put to-

[5]Ibid., 174.

gether winning hands much more often and the odds of starving to death dropped.

Consider the story of American food crops in a land usually thought of as resistant to outside influences: China. No Old World people adopted these alien plants faster than the Chinese. The explanation for their eagerness probably is related to population pressure. Between 1368 and 1644, the years of the Ming dynasty, the Chinese population doubled, and so did the demand for food. Farmers of the traditional staples, such as wheat in the north and rice in the south, were running into problems of diminishing returns. They were close to raising as much food as they possibly could on available land using existing techniques. The problem must have been especially pressing in the south, where most of the level and near-level land close to sources of water for irrigation was already occupied by rice paddies. To return to the poker analogy, the Chinese needed to draw more cards.

The Iberians, Portuguese and Spanish, provided the new cards. The port of Manila, newly Spanish and an easy sail from the China coast, was particularly significant in the transfer of Native American crops to China. The sweet potato, a calorically rich food, finally made it all the way across the Pacific and arrived on the China coast some time in the last years of the sixteenth century. This crop prospers in inferior soils, tolerates drought, resists insect pests, and does well with very little attention, so the peasants raised rice to sell and to pay taxes, and sweet potatoes to eat. By 1650, the crop was common in Guangdong and Fujian provinces, and on its way to becoming the staple of the poorer peasant wherever climate would allow. Today, China raises more sweet potatoes by far than any other nation.

Maize arrived in China even before the mid-sixteenth century. It too was hardy and required little attention. It produced food faster than most crops and provided high yields in the relatively cool and dry uplands of the interior. Maize soon became a common secondary crop from Shanxi in the northwest to Yunnan in the southwest and eventually a primary crop in several inland provinces. Today, China is second only to the United States in maize production and, unlike America, devotes almost all of it to feeding humans, not animals.

Farmers and gardeners were slower to adopt American cultivars at the other end of Eurasia. Not until the eighteenth century did the peasants of the northern half of Europe from the British Isles to Poland, an area suited in soil and climate for potatoes, plant the tubers in massive quantities. Next the Russians, for whom a day without at least one meal including potatoes is still a rarity, fell in line. It is questionable whether Northern Europe could have supported its population growth after 1750 or fed the laborers of the first century of its industrial revolution without potatoes. In the northerly latitudes of Europe, where summers were too short and wet for wheat, the farmers had chiefly depended on rye, a nourishing but not very productive grain. Potatoes thrived there, producing four times more calories per acre than rye. Even better, they could be planted in fields lying fallow between rye plantings, converting seasons of zero food production in said fields to seasons of plenty. Maize played much the same role in northern Spain and Portugal and southern France, where in the 1670s John Locke saw fields of it, called *blé d'Espagne* locally, where it "serves the poor people for bred."[6] Between 1900 and 1914, the annual consumption of potatoes per capita in the United Kingdom was 97 kilograms; in France, 176; in Germany, 199; and in Belgium, 206.5. Even the Italians, for all their affection for maize, consumed 27.8 kilograms.

According to the demographic historian Ping-ti Ho, writing about China, "During the last two centuries when rice culture was gradually approaching its limit, and encountering the law of diminishing returns, the various . . . food crops introduced from America have contributed most to the increase in national food production

---

[6]Crosby, *The Columbian Exchange*, 178.

and have made possible a continual growth in population."[7] His statement can be appropriately extended to the entire Old World. His concept, if augmented with a sentence pointing to the influence of Old World plants and animals in the New World, accurately describes the influence of the Columbian Exchange on the entire population of the globe. Now humanity could exploit the full potential of all its Neolithic agricultures.

CODA: THE IRISH EXCEPTION AND THE POTATO FAMINE

For the human species as a whole, sharing domesticated plants and animals, that is, combining the hands nature had dealt them separately in the Old World and the New, enriched lives and stimulated population growth. There were, however, exceptions. Ireland, sodden and cool, was the first European country in which potatoes became an essential food. That happened in the seventeenth century when the English broke Irish resistance and pushed the natives off the best land, making tenant farmers of many of them. The Irish had to resort to celibacy and infanticide or find some way to raise more food on less land. They chose to cultivate potatoes.

Irish peasants discovered that they could plant and harvest the tubers with the simplest of homemade gear and little labor, and that they could extract enough nourishment from the marginal plots allowed them by their landlords to live and multiply. An acre and a half produced enough potatoes to, with a little milk, support a whole family. It was not exceptional for an adult to consume 10 pounds of potatoes a day. On such a healthy if precarious diet, Ire-

[7]Ping-ti Ho, *Studies on the Population of China* (Cambridge, MA: Harvard University Press, 1959), 191–92.

land's population rose from 3.2 million in 1754 to nearly 8.2 million in 1845. Two fifths of them, over 3.0 million, relied on the potato as their staple. A third of Ireland's farmland was planted with that tuber.

In the mid-1840s an American fungus, *Phytophthora infestans*, arrived in Europe and spread, turning potatoes into black rot all across the continent, rendering the lower classes miserable everywhere the potato had become an important part of diet. But only in Ireland had the potato become the irreplaceable staple, and only there did tragedy ensue.

In the years 1846–51 approximately a million Irish, at that time citizens of Great Britain, foremost beneficiary of the expansion of Europe across the oceans and of the burgeoning industrial revolution, died of starvation and associated diseases. A million and a half more emigrated to other countries. The societies of Western Europe could not reconcile the horror of the Potato Famine with their pride in the humanitarian advances of an era that saw the end of the Atlantic slave trade and, in a few more years, of slavery itself throughout the Atlantic world. So for a century the Great Hunger was omitted from consideration or, so to speak, reduced to an appendix at the back of the book. The population of Ireland (including Northern Ireland) has never regained its pre-famine level of over 8 million.

The significance of the famine is best measured quantitatively—so many deaths here, so many immigrants there—but we shouldn't let the numbers distract us from the realization that this was something that happened to individual human beings. Consider William Ford, who in 1847 emigrated from County Cork to America and found his way to Dearborn, Michigan, via oxcart and the Erie Canal. There he built a farmstead out of what a generation before had been frontier forest. He married Mary O'Hern and in 1863 the couple had a son, Henry Ford, who found humanity on horseback and who would eventually put humanity into vehicles powered by a new prime mover burning fossilized sunlight.

PART TWO

FOSSILIZED SUNSHINE

Man's function as a force of nature was to assimilate other forces as he assimilated food.

—Henry Adams, historian (1906)[1]

Collectively, heat engines are the motive force behind our civilization.

—David Goodstein, physicist (2004)[2]

For nearly all of our time on earth, we have been at least as feeble as other animals in the physical power to get work done. When Benjamin Franklin, whose discoveries in the field of electricity would prove so significant in energy distribution, was born in 1706, we humans were only marginally more powerful than we had been in the age of the Egyptian pyramid builders. Since then we had added water and windmills as prime movers (more about them in a few pages), but to accomplish most tasks—plow a field, harpoon a whale—humans still used muscle as their prime mover and food for fuel.

In 1706, we used horses, oxen, donkeys, water buffalo, and other domesticated animals to do some of the work, for example, hitching them up to millstones to walk in circles to grind grain. We often, very often, used people for the same purposes, by setting some to

[1] Henry Adams, *The Education of Henry Adams* (New York: Penguin Books, 1995), 449.
[2] David Goodstein, *Out of Gas: The End of the Age of Oil* (New York: W. W. Norton & Company, 2004), 88.

plodding on treadmills to lift water in chains of buckets, and others to turning big hollow wheels from inside like hamsters in order to saw wood, drive bellows, knead dough, and so on. The eighteenth was the peak century of the slave trade because Europeans needed more muscle than their home continent could or would supply to exploit their American plantations, so they went to Africa to get it.

In that century an early pioneer in the development of a steam engine, John Smeaton, estimated that in prolonged efforts, humans produce (to express it in our terminology) about 90–100 watts of power apiece. A plantation owner or entrepreneur might try to compensate for such feebleness by utilizing more people, but there was no way to get a sustained 500 watts of power per individual even with a whip or improved diet.

Humanity had hit a ceiling in its utilization of sun energy. The combined efforts of a million slaves could not get a pound of West Indian or Brazilian sugar to Europe faster than a sailing ship. The labor of a million Roman candlemakers could not produce enough light to permit the staging of night games in their city's Colosseum. Humanity could produce Islam and Buddhism, algebra and calculus, Shakespeare and Palestrina, but it couldn't produce decent shoes for more than a minority of those who needed them, or enough type, paper, and ink for a single daily newspaper with a circulation of a million, or a means to get from Rome to Paris in less than a week. Humanity required an energy miracle commensurate with its domestication of fire if it was to become the bully of the biosphere.

By the time that Benjamin Franklin died in 1790, humanity had its miracle, a previously neglected kind of fuel and a new prime mover. The industrial revolution had been ignited, bringing with it an unprecedented acceleration of technological innovation, economic and political rearrangement and consolidation, and demographic increase and migration. The new means of exploiting sunlight radically altered the lives of the citizens of the English Midlands, then of the Ruhr, and, a few generations later, those of the inhabitants of Mongolia, the South Seas, the headwaters of the

Amazon, and of everyone else except a tiny and dwindling number of the most remote hunter-gatherers.

We will deal with the new prime movers one at a time in later chapters in the order that they arrived on the scene. They were heat engines, direct descendants of the pots that had been boiling over hearths for a very long time. They were machines that tapped sunlight by burning its manifestations in biomass. The difference in the fuels burning on most hearths and those burning in the new prime movers was in their energy density and in their age.

The energy density of recently cut wood or foliage or grass is slight per unit of weight or volume. A solution to that problem is to concentrate biomass: to somehow reduce an inconveniently bulky ton of wood to a neat pound of fuel with the same energy potential. Eighteenth-century people all over the world did something like that when they burned wood slowly in nearly airless conditions to make it into charcoal; but the reduction was insufficient and there weren't enough trees to supply enough wood for very long—and charcoal wasn't suitable for all or even most of the tasks humans wanted heat engines to accomplish.

Humans blindly turned to the greatest reductionist of all: time. Life has existed on our planet for eons, during which billions of generations of plants and animals have lived and died and left their corpses behind. Chance buried a fraction of these, and a fraction of this fraction has been subjected to conditions that transformed it into the fuels that empower our civilization: peat, coal, oil, and natural gas.

Peat—a light, spongy combustible—is what we might call the preliminary, the adolescent fossil fuel. It is vegetable matter, usually no more than a few thousand years old, that accumulated in stagnant bogs where conditions restricted the activity of bacteria and minimized the amount of oxygen, slowing decomposition. Its energy density is higher than that of fresh biomass, but still low; it burns smokily; and it smells. Its existence in enormous and easily accessible beds in many parts of the world and its low price are its recommendations.

Coal is what can happen to peat after a long time. It is the reduced and compacted remains of forests millions upon millions of years old, forests of club mosses, some 30 meters high, of giant horsetail rushes, of flora in a profusion barely equaled by our most luxuriant jungles today. Its energy density is at the very least triple that of peat.

Oil is what we have left of tiny organisms, phytoplankton, that lived in oceans long, long ago, died and accumulated in oxygen-deficient waters, and then were buried 7,500 to 15,000 feet deep long enough for pressure and heat to transform them into liquid. (Below 15,000 feet the conditions are so extreme that gas, not liquid, is the result.) This liquid then collected beneath impermeable caps of stone to await our pleasure. The energy density of oil is a great deal higher than that of coal and it is liquid and therefore easier to store and transport.

Fossil fuels are the tiny residue of immense quantities of plant matter. An American gallon of gasoline corresponds to about 90 tons of plant matter, the equivalent of 40 acres of wheat—seeds, roots, stalks, and all. Coal, oil, and natural gas are the end products of an immensity of exploitation of sunshine via photosynthesis over periods of time measured by the same calendars used for the tectonic shuffling of continental plates. We are living off a bequest of fossil fuel from epochs before there were humans and even before there were dinosaurs.

4

COAL AND STEAM

And was Jerusalem builded here
Among these dark Satanic Mills?
—*William Blake, poet (1804)*[1]

The world is now entering upon the mechanical epoch. There is nothing in the future more sure than the great triumphs which that epoch is to achieve. It has already advanced to some glorious conquests. What miracles of invention now crowd upon us! Look abroad, and contemplate the infinite achievements of the steam power.

—*Robert H. Thurston, historian (1878)*[2]

Accident and evolution gifted humanity with culture, which made the species into a sprinter. Humanity had dashed ahead twice before, once in the Upper Paleolithic and again in the Neolithic. In the eighteenth century, humanity sprang out of the blocks the third time. The surge of mechanization that started in England was unprecedented in the speed of its advance and its general influence, but it wasn't all new in its basic physical materials. Wood still sufficed for many purposes; indeed, it still does in our developing nations, especially for cooking and warmth.

---

[1] *The Harper Anthology of Poetry*, ed. John F. Nims (New York: Harper & Row, 1981), 244.

[2] Robert H. Thurston, *The History of the Growth of the Steam Engine* (New York: D. Appleton & Co., 1878). Available at http://www.history .rochester.edu/steam/thurston/1878/. (Viewed June 3, 2003.)

Metals—copper, bronze, iron, steel—were expensive but available for tasks requiring special strength, hardness, and durability. The machines of this industrial revolution were new, but their components and basic concepts—wheels, levers, pulleys, screws, and so on—were of ancient lineage.

One of these parts was the wheel, which dates from about 5500 BP. A millennium or so later, Old World peoples had wonderfully useful wheel machines: wagons, chariots, potter's wheels, and others. The proof of their significance is the relative disadvantage of the societies that lacked the wheel, those, for instance, of the high civilizations of America. How often and for how long has a people without the wheel ever subjugated a people with the wheel?

In the first century AD, humans invented a complement to the wheel that became so common that it is hard to imagine life without it: the crank. It is a bar or rod attached to a shaft at a right angle, usually with a handle parallel to the shaft jutting out at the far end for the convenience of anyone who wants to turn or "crank" (the word is a verb as well as a noun) the shaft. With it, members of my grandparents' generation started the engines of Model-T Fords. Today, when I want to sharpen my pencil, I insert it into the pencil sharpener and turn . . . the crank.

The peoples who possessed and fell in love with the wheel and crank—the Chinese and Europeans, for instance—were using them to squeeze more energy out of sunlight long before the eighteenth century by exploiting the movement of water and wind.

The watermill, an ancient prime mover churning away on the Tigris and other Old World streams before Jesus or Mohammed, taps the potential energy stored in water deposited in ponds, lakes, snowpack, and glaciers at high elevations by sun-driven evaporation. The water makes that energy available as it flows back downhill toward the sea.[3] Watermills were at first, and for long after in

---

[3]Tidal mills tap lunar, not solar, power; they are omitted for that reason and because they have been of major importance in very few locations.

Mechanism of a watermill, engraving from V. Zonca's
Nova teatro di machine, 1607.

mid- and eastern Asia, structured with vertical axes—like revolving doors set up in streams. Compared to what followed, they were inefficient because the flowing water pushing the flaps or vanes in one direction was also, unless carefully diverted or otherwise restrained, pushing against the flaps revolving around the axis in the other direction. Some inventive soul—who, when, and where we don't know—shifted the wheel's axis to the horizontal and set up the wheel so as to dip into the moving water from above. The flow pushed the wheel in only one direction. This is the undershot wheel, illustrated on page 65.

It was an improvement, but worked well only on swiftly flowing streams. Then humans built flumes or chutes that brought water from higher elevations to the top of the wheel, where it poured into troughs that were the vanes of this new edition of the old wheel. This tapped the energy not only of water flow but of gravity, the weight and impact of the falling water. A well-built, well-lubricated "overshot wheel," as this kind is called, might produce four or five or even more horsepower. According to Fernand Braudel, the great French historian, an average watermill ground five times as much grain as two men working a hand mill. In the early twelfth century, France alone had twenty thousand watermills grinding wheat, crushing ore, and so on, the energy equivalent of a half million human workers.

In that century waterwheels were common along rivers and lesser streams right across Eurasia and North Africa from the Atlantic to the Pacific. Since then they have been built throughout the rest of the world, where many are still operating, especially in the Third World. If we include in our definition of watermills the giant turbines paired to our dams today that drive dynamos to produce electricity, then this prime mover is as important today as it ever was.

Mills constructed to exploit sun-driven wind—windmills, which first appeared in Persia in the first millennium CE—also started out like revolving doors. That kind were common right up to our time throughout Asia, where the demand for power exceeded what muscle and water could provide; but someone somewhere, probably in Western Europe in the twelfth century, invented a better windmill. He made a giant vertical fan of the revolving doors, faced it into the wind, and greatly improved the machine's efficiency, bequeathing us what we see in our minds when we think of windmills.

Windmills were and are very useful in windy regions like Southeast Iran, the coasts of China, and, famously, the Netherlands. In the latter in 1650 there were at least eight thousand windmills towering over the sodden countryside pumping water. Well into the second half of the twentieth century windmills were standard equipment in American farms, especially in the semi-arid plains. More and more of them are being raised today because windmills don't pollute.

But flowing water and wind were far from efficient expressions of sun energy. The first froze in winter and sometimes ran low in summer and, anyway, was not available everywhere. The second sometimes blew too hard and sometimes not at all. In the long run, the significance of the mills may be that the people who built and maintained them learned a great deal about levers, axles, cogs, pulleys, screws, and so on. Before the human species could make another quantum jump in exploitation of sun energy, it would have to gain access to much more energy than food or water or wind could supply; it would have to find a better fuel and then invent a more powerful prime mover.

Humanity lusted after concentrated sun energy, of which there were portents. The Byzantines invented Greek fire, a flammable liquid that stuck to everything and could be squirted short distances. Water didn't put it out and so it was very useful in naval battles. The Greeks kept the recipe for Greek fire secret, but it must have included the fossilized sun energy, petroleum, that seeped to the Earth's surface in many places in the Balkans and Middle East. The Chinese invented gunpowder, an extreme kind of fuel. At first it was used as an elixir, then in a sort of flamethrower, then in bombs to unleash a shrapnel of feces and broken crockery among the enemy. Not long after 2000 BP they were using it in guns. The ignition of a fuel—in this case, gunpowder—in a closed container produced enormous push, flinging missiles at velocities so swift that the human eye couldn't follow them. The cannon barrel and cannon ball were prototypes for the piston and cylinder.

Humans did not consider coal a fuel when they first met up with it, but valued it for its peculiar color and even used it in jewelry. In time they learned that it would burn, and a few were using it as a source of heat and light thousands of years ago. But accessible surface outcroppings of the fuel existed in only a few places, and a lot of that coal produced choking, eye-watering smoke. And wood was plentiful, at least to begin with, so there was no need to suffer coal for energy.

There were at least two false or, if you prefer, preliminary starts on revolutionizing industrial production, one in eastern and one in western Eurasia. The ironmongers and miners of China's Song dynasty started an industrial revolution of their own seven or eight hundred years before the Western Europeans got around to theirs. In 1078, China used huge quantities of charcoal to process ore into 125,000 tons of iron, twice as much as Europe (excluding Russia) produced four hundred years later. But then their revolution faltered as they ran into shortages of wood. These ironmongers began to switch to coal, of which China has large deposits in the north and northwest. That solved half their problem; but they did not tap the full sun energy of coal, which would have required inventing a new prime mover. Perhaps that was simply a matter of chance, which plays a more important role in history than many historians like to admit. Perhaps the explanation lies in the disasters that rolled over northern China: barbarian invasions, vicious civil wars, Yellow River floods. China's political and economic center of gravity shifted south and away from its richest coal deposits. Coal was not widely adopted as a substitute fuel for biomass in China until the twentieth century.

In the 1600s, when New York was born as New Amsterdam, and Cape Town and Malacca were also Dutch cities, Holland ran out of harvestable wood and couldn't shift to coal unless it imported it. The Dutch shifted to peat, a fossil fuel of which they had plenty. They warmed their homes, cooked their meals, and processed many of their manufactured products—brewed their beers, refined their sugar, baked their bricks—with peat fires. Rembrandt van Rijn painted his masterpieces as a citizen of the first society primarily powered by fossil fuels, arguably the first modern society. But, like the Chinese, the Dutch did not come up with a revolutionary prime mover. They started down the path that led to fossil fuel civilization, but halted halfway because peat, which fulfilled their immediate needs for fuel, didn't burn hot enough per unit of volume or weight to entice them on to invent a new prime mover.

The birthplace of the lasting, possibly perpetual industrial revolution turned out to be Great Britain. Many explanations have been suggested: a sturdy artisan class, Protestant discipline, an excellent transportation system of rivers and coastal waters, a market freer than most, a relatively dependable currency, relatively honest bankers, and so on, but these were characteristics of Dutch society as well. Whatever the cause or causes of Britain's industrial revolution, the presence of enormous quantities of coal under its soil was an essential ingredient for that revolution. The stimulus for the switch from biomass to coal as the primary source of sun energy was simply that England, Wales, and Scotland, like China and the Netherlands, were running short of forests. The price of firewood in Britain rose 700 percent between 1500 and 1630, much faster than general inflation. In 1608, a census of the number of "tymber trees" in seven of Britain's largest forests set the sum at 232,011. The number in 1783 was down to 51,500. The British imported wood from the Baltics and North America, but their chief solution to their fuel problem was to mine more and more coal.

At the end of the seventeenth century, London was already famous for its smoke. The diarist and gardener John Evelyn, as exas-

perated as a Los Angeles environmentalist on a bad smog day, com-
plained that London resembled "Mount Aetna, the Court of Vul-
can, Stromboli, or the Suburbs of Hell, rather than an Assembly of
Rational Creatures, and the Imperial Seat of our incomparable
Monarch."[4] Most of the city's coal came by water from Newcastle
up the coast, hence the mysterious old saw about the inappropriate-
ness of bringing coals to Newcastle.

Coal was plentiful in the English homeland, but surface outcroppings were used up fast, and by 1700 mine shafts were as deep
as 200 feet. There were problems down there with gases and especially with flooding. Dig a 200-foot hole in a countryside with high
rainfall, and even if you don't hit springs, the bottom of the hole
will fill up with water. The miners tried lining the walls of mine
shafts with sheep skins to hold back the water, but that didn't help
at all. They dug tunnels to drain the water out of the mines, but this
only worked when the shafts were in the side of a hill and gravity
could be enlisted to carry the water down into the valleys. Muscle,
animal and human, and sometimes watermills and windmills were
put to work lifting the water out of the mines, but it was an endless
battle that technology circa 1700 could not win. Britain's industrial
revolution was drowning *in utero*.

Coal, the Carboniferous legacy of stored sunlight, would solve
that problem. Coal would be burned to power the *heat engine*, which
my desk encyclopedia defines as "a device that transforms disor-
dered heat activity into ordered, useful, mechanical work."[5] The
story of that kind of contrivance, of which today's nuclear reactors
are examples, begins with the eighteenth century's steam engines.

Humans had known about the power of steam for as long as
they had boiled water and watched the pot lid tremble and lift. In

[4]Quoted in Barbara Freese, *Coal: A Human History* (Cambridge, MA:
Perseus Publishing, 2003), 35.
[5]*The Cambridge Encyclopedia*, ed. David Crystal (Cambridge: Cambridge University Press, 1990), 554.

the first century AD the proto-scientist, Hero of Alexandria, made what we would describe as a sort of lawn sprinkler arrangement powered by steam and watched it spin. Yes, humanity knew that the expansion of steam could provide power, but it was a long walk from that realization to a practical steam engine.

There were enormous difficulties. There were no smiths who made parts in accordance with precise measurements. Metals resistant to high heat and strong enough and light enough to be used for boilers and cylinders, which would contain and restrain steam under pressure, weren't available in quantity. There had never been a useful machine combining piston and cylinder, unless you count the cannon ball and cannon as such. Inventing the first practical steam engine would have to be an event on the far frontier of possibilities.

The key concept could not be to use steam to *drive* pistons vigorously and rapidly because machines with such capabilities were barely imaginable. The way around that—to us an odd one—was to use steam to make a vacuum. Classical and medieval Europeans did not believe in vacuum: space existed because there was something in it, even if only air. But halfway through the seventeenth century a German, Otto von Guericke, pumped the air out of a sphere of two carefully fitted halves, and then sixteen horses couldn't pull them apart. In 1680, a Dutchman, Christian Huygens, suggested that exploding gunpowder under a piston in a cylinder would drive out the air and most of its own gases. What would be left would be nearly nothing, at least a partial vacuum. Then atmospheric pressure would push the piston down into the vacuum. Some bright soul might try to harness that motion to do work.

The man who created that engine was an Englishman, Thomas Newcomen, of whom we know very little beyond the dates of his birth and death, 1663 and 1729. We don't even know where he was buried. He was an ironmonger from Dartmouth, one of the robust artisans produced by Western Europe's growing economies. He had little if any formal education (in that he reminds us of Michael Fara-

day, who figures importantly in the chapter on electricity). Perhaps
he had read Huygens and the like, but probably not. He was a Bap-
tist, a Nonconformist in religion like so many of the pioneers of
Britain's industrial revolution. His admiring biographers identify
him as the "first great mechanical engineer."[6]

Today, it is easier for us to imagine using hydrogen fusion to
heat our homes than for Englishmen circa 1700 to imagine using
fire to pump water out of mines. You set things on fire to destroy
them or for warmth and light, not to push or pull, and Europeans
didn't even have good fireplaces or stoves. Those that they had let
most of the hot gases rise up the chimney to heat all outdoors. The
Franklin (Benjamin Franklin) Stove, which detained the gases and
tapped their heat before releasing them, didn't come along for more
than a dozen years after Newcomen's death.

We are too accustomed to energy in plenty to comprehend the
energy poverty of the world that Newcomen lived in. To illustrate:
In 1682 in obedience to the order of Louis XIV, the "Sun King,"
fourteen waterwheels 12 meters in diameter were built near Ver-
sailles to harness the currents of the Seine to pump water to its
fountains. They were together called "the eighth wonder of the
world," but delivered at the very best 124 and usually not more than
75 horsepower. We scorn automobile engines of such feebleness
today; three centuries ago an engine that small and of such power
would have been worshipped.

Newcomen, with the help of a plumber named John Cawley or
Calley, built a steam machine close by a coal shaft 51 yards deep at
Dudley Castle, Staffordshire, in 1712 (see illustration on page 73).
He didn't use gunpowder explosions to produce vacuums and move
pistons, but steam. The machine's boiler held 673 gallons of water.
Its cylinder, vertical, was 21 inches in diameter and 7 feet 10 inches
tall. The fit of the piston and cylinder was loose and had to be sealed

[6]L. T. C. Rolt and J. S. Allen, *The Steam Engine of Thomas Newcomen*
(New York: Science History Publications, 1977), 13.

Newcomen Engine, as modified by Richard Trevithick at Bullan Garden, Dolcoath, from *Life of Trevithick*, 1775.

with a wet leather disk. Steam from the coal-heated boiler admitted into the cylinder lifted the piston up. Then cold water sprayed into the interior of the cylinder. The voluminous steam condensed to a small amount of liquid, producing a vacuum in the cylinder. *Then* came the power stroke, for which all the above was preparatory, as atmospheric pressure drove the piston down into the evacuated cylinder. The piston was attached to a rocker beam, the motion of which could drive a chain of buckets, a bellows, and so on.

Newcomen's first machine made twelve strokes a minute, raising 10 gallons of water with each stroke. Its strength is estimated at 5.5 horsepower, not impressive to us, but the "fire engine," as it was sometimes called, was a sensation in power-starved Britain and Europe. Soon there were scores of Newcomen engines, most nodding at the pitheads of Britain's mines, which now could be dug twice as deep as before. In 1700, Britain produced 2.7 million metric tons of coal; in 1815, 23 million tons. That sum was twenty times in energy equivalent what the existing woodlands of Britain could produce in a year. If that quantity of coal were burned in steam engines (a lot of it, of course, went for heat and cooking), the amount of power created would have equaled that of 50 million men.

At least fifteen hundred Newcomen machines were built in the eighteenth century. The rapidity of their spread in an age still, by our standards, more medieval than modern provides a measure of the need they answered. The first Newcomen engine on the continent was constructed at Königsberg in 1722. When Newcomen died in 1729, his engines were operating in Saxony, France, Belgium, and perhaps Spain. The first Newcomen engine in the New World was built in 1753 at the juncture of Belville and Schuyler Avenues in North Arlington, New Jersey. In 1775, John Smeaton built a Newcomen engine to drain Kronstadt's great drydock, which two large windmills were not keeping dry enough. This had a cylinder 5 foot 6 inches in diameter.

Thomas Newcomen's invention was the first machine to provide significantly large amounts of power not derived from muscle, water, or wind. It was a new prime mover. It utilized fire—a natural force—to heat water, to make steam, to do work. It was the first practical machine to use a piston in a cylinder. It worked night and day. If I were to attempt anything so simple-minded as to pick a birthday for the industrial revolution, it would be the first day that Newcomen's engine began operating in 1712.

In the eighteenth century, the Newcomen machine saved Britain's coal industry if not from watery demise then certainly from soggy stagnation, enabling Britain to continue with industrialization. Without increasing supplies of fossilized and concentrated sun energy, Britain's industrial revolution would have fizzled like China's. Without Newcomen's invention there would have been many fewer people in the eighteenth century thinking in terms of heat engines. Without Newcomen's invention James Watt, who for some reason is commonly credited as *the* inventor of the steam engine, actually would have had to invent the first one.

The Newcomen engine was grossly inefficient, as beginnings usually are. It was an impressive prime mover compared to muscle, water-, and windmills, but was wildly wasteful of fuel. It heated water into steam, introduced that into a cylinder, and then sprayed water into the cylinder. An egregious proportion of the fuel's energy was wasted reheating the cold cylinder at every stroke. Most of the Newcomen engines stood at the pitheads of coal mines where coal was cheaper than elsewhere, but even there it was not quite free. Big Newcomen engines consumed coal to the value of £3,000 a year, a hefty sum.

Moreover, as we, with our advantage of hindsight, can proclaim, the Newcomen engine was too big and awkward to be adapted for many tasks. It might possibly do for large ships, but not for land transportation. One at least was built to travel about on a wagon, but it was in the same category as Dr. Johnson's dog that could walk on its hind legs—interesting, but not really practical.

In addition, there was a problem with the kind of motion it supplied. Most of the machines of the budding industrial revolution required rapid, steady, and rotary motion. The Newcomen rocker arm moved slowly, not always evenly, and up and down, not round and round. When smoothly flowing rotary action was required, mill owners used Newcomen engines to raise water to pour down on waterwheels, and then the wheels did the work.

The Newcomen engine was the only practical steam engine for sixty or seventy years. Then a generation of mechanical engineers matured who were familiar with its defects, acquainted with the sciences, and had access to better materials, tools, and more skillful craftsmen than Newcomen had ever had. For instance, after 1774, British engineers who wanted pistons of precise measurements could get them from a new machine for boring cannon barrels.

The first hero of the new generation was James Watt, a Scot, a Presbyterian, a Nonconformist like Newcomen, well educated, and friendly with scientists and venture capitalists. The most important of the latter was Matthew Boulton, who wrote to Watt that he had two motives for investing in his project: their friendship and his "love of a money-getting, ingenious project. . . ." Watt was repairing a model of a Newcomen engine in 1764 when it occurred to him that the engine wasted fuel reheating the cylinder at every stroke. Instead of spraying water into the hot cylinder, why not let the pressure of the steam and the descending piston push the steam into an adjoining and unheated chamber to condense?

A dozen years later he installed two such engines, one to pump water and another to blow air into blast furnaces. They functioned satisfactorily, and by 1790 Watt had joined Boulton in the wealthy elite. By 1800, the Watt engine was producing three times as much power per bushel of coal as even the latest models of the Newcomen engine, and was pumping water out of mines, turning out flour, paper, iron, and so on and on. In the years around the turn of the nineteenth century, after Watt's patent ran out, improvements and applications of his engine came in a rush. The original version consumed too much coal and was too big and heavy to be a practical power source on vehicles or ships. It was also restricted in power and speed by its Newcomen-like dependence on atmospheric pressure to drive the piston into the cylinder as the steam was being expelled. Moreover, the arrangements Watt provided to transform the piston's linear motion to circular motion were more ingenious than effective. Within a few years high-pressure steam engines with pistons driven both back and forth by steam and linked to wheels by cranks and connecting rods were revolutionizing British activities at home and, a few years later, elsewhere.

The magnitude of the influence of the steam engine was clearly apparent in transportation. Primitive locomotives on rails began puffing along early in the nineteenth century. In 1830, one called the *Rocket* pulled a train from Liverpool to Manchester. A decade later, Britain had 1,400 miles of railroad track, continental Europe 1,500 miles, and the United States, sprawled across a continent, 4,600. In 1869, the United States, by then a nation with two populated coasts and a mostly vacant middle, tied the coasts together with the first transcontinental railroad. Elsewhere men with political influence and access to cartloads of capital were thinking about a Cape-to-Cairo railroad, a Trans-Siberian railroad, and other such awesome ventures.

The revolution that steam enacted on the oceans was at least as spectacular. In 1838, the paddle wheelers *Sirius* and *Great Western* raced from British ports to New York City for the honor of being the first vessel to make the westward crossing by steam only. The winner was the *Sirius*, but only because it set sail (an anachronistic verb we still use) ten days before the *Great Western*. The *Sirius* ran out of coal off New Jersey and burned cabin furniture, all the spare yards, and one mast under the boiler of its 320-horsepower steam engine to get to New York four hours before its rival. The *Sirius* made the crossing in eighteen days and ten hours at an average speed of 6.7 knots an hour. The *Great Western* averaged 8 knots an hour and completed the passage in fifteen days. It had four boilers, two engines of the latest design, and arrived in New York with 200 tons of coal in its bunkers. Both ships crossed the Atlantic in about half the time usually taken by ships dependent on wind.

ROBERT H. THURSTON, one of the first historians of the steam engine, opined in 1878 that Watt's hometown, Greenock, originally a fishing village, now "launches upon the water of the Clyde a fleet of steamships whose engines are probably, in the aggregate, far more powerful than were all the engines in the world on the date of Watt's birth, January 19, 1736."[7] But shipping was merely one spectacular manifestation of the impact of Watt's innovation; its influence was greatest in the textile industry, the first to be mechanized. As early as 1800, the steam-driven spinning mule (a machine for spinning cotton thread and winding it on spindles) could produce as much per unit of time as two hundred to three hundred human spinners.

---

[7] http://www.history.rochester.edu/steam/thurston/1878/ chapter 3, n.p.

The production of manufactured articles rose exponentially, as did trade, internal and worldwide. Between 1771 and 1775, England imported 5 million pounds of raw cotton; in 1841, 58 million pounds. In 1834, it exported 556 million yards of woven cotton goods. In that year, over 8 million "mules" were at work in England's cotton mills, which employed 220,000 workers. Most of the power in the cotton mills was supplied not by muscle but by the newer prime movers: 11,000 horsepower by waterwheels and 33,000 horsepower by steam engines. The factories of England and, soon after, New England turned out thousands of miles of cotton thread and cloth annually, making more inexpensive and decent clothing available than ever before. The ramifications of this textile explosion were not all positive, however. Plantations in the southern United States shifted to a monoculture of cotton cultivation, a labor-intensive endeavor. And the power there was still supplied by the primitive prime mover of muscle, in the form of slavery, an institution whose slumping status was revived by the profits to be made from the new textile mills.

The industrial (i.e., steam) revolution of the nineteenth century radically altered the global economy and thereby the global balance of power. For instance, the speed and output of England's mills blighted and nearly extinguished the ancient textile industry of India, disrupting the lives of thousands there. During the eighteenth century, India, China, and Europe had accounted for about 70 percent of the world's gross domestic product, each providing roughly (very roughly) one third. By 1900, China's share of the world's manufacturing output was down to 7 percent, India's to 2 percent; Europe's was 60 percent and the United States's 20 percent.

Steam immensely enhanced the speed and dependability of transportation, and spurred migration within countries—from rural to urban, from settled to frontier—often via railroad. Myriads of migrants left Europe for its overseas colonies and the United States: 400,000 people a year from 1850 to 1900, and then 1 million a year from 1900 to 1914. It was as if the shift to steam had pumped

Europe full and it was exploding and flinging fragments of itself over the oceans. In the same decades millions left India and China for the Americas, South and East Africa, Mauritius, the Pacific Islands, and elsewhere to work in plantations, to build docks, roads, and railroads. Many sent their wages back home to their families and eventually rejoined them there; others brought their families to join them overseas, and yet others acquired spouses to found new families in the new lands. Whatever the origins of the migrants, the vast majority crossed the oceans by steamship. The total number of all the ocean-crossing migrants between 1830 and 1914 was an amazing 100 million.

Nothing like this colossal increase in productivity, shifts in global power and reach, alterations in locus of global hegemony, and movement of goods and people had ever happened before. Taking the measure of the steam engine's gravitas, of its historical *mass*, using the word as physicists do, is like trying to judge the significance of the tectonic drift of the continents. Let us consult expert witnesses.

Friedrich Engels was an honest witness of the human costs of the early industrial revolution. He was appalled by the horrors that followed when English factory owners obliged laborers to work to the tempo, mechanical and economic, of the industrial revolution: the illnesses and accidents, the malformations of bones and joints and minds dictated by repetitive labor, the hunger, the insecurity, the general wear and tear that shortened lives. A good proto-sociologist (though occasionally succumbing to the temptation of the exclamation mark and often to that of making prophecies), he collected statistics on England's factory workers circa 1840:

> Of 1,600 operatives employed in several factories in Harpur and Lanark, but 10 were over 45 years of age; of 22,094 operatives in diverse factories in Stockport and Manchester, but 143 were over 45 years old. Of these 143, 16 were retained as a special favour, and one was doing the work of a child. . . . Of

fifty worked-out spinners in Bolton only two were over 50 and the rest did not average 40 and all were without means of support by reason of old age! . . .

In all directions, whithersoever we may turn, we find want and disease permanent or temporary, and demoralization arising from the condition of the workers; in all directions slow but sure undermining, and final destruction of the human being physically as well as mentally.

Engels predicted that soon the day would come when the proletariat would rise and then "will the war cry resound through the land: 'War to the palaces, peace to the cottages!'—but then it will be too late for the rich to beware."[8]

Charles Dickens was also appalled by the poverty, hunger, disease, and anguish of the peasants who migrated from countryside to city to work in the new factories and, as well, by the ever-thickening coal smoke of the industrializing cities. In his novel *Hard Times*, he provides a description of Coketown, his fictional but accurately representative city of the early industrial revolution:

It was a town of machinery and tall chimneys, out of which interminable serpents of smoke trailed themselves for ever and ever, and never got uncoiled. It had a black canal in it, and a river that ran purple with ill-smelling dye, and vast piles of buildings full of windows where there was a rattling and a trembling all day long, and where the piston of the steam-engine worked monotonously up and down like the head of an elephant in a state of melancholy madness.[9]

[8]Friedrich Engels, *The Condition of the Working Class in England* (New York: Viking/Penguin, 1987), 179, 221, 292.

[9]Charles Dickens, *Hard Times* (London: Folio Society, 1994), Book I, chapter V, p. 18.

Dickens also noted, grudgingly and as an afterthought, that the new mills produced "comforts of life which found their way all over the world."[10] He saw both sides of the coin of industrialism, withheld predictions, and contented himself with outrage.

The American statesman and advocate for American industry Daniel Webster was in 1818 much more single-minded than Dickens. He focused on the positive characteristics of the steam engine: its power, adaptability, and promise. To the members of the Boston Mechanics' Institution, he extolled the virtues of the steam engine:

> It rows, it pumps, it excavates, it carries, it draws, it lifts, it hammers, it weaves, it prints. It seems to say to men, at least to the class of artisans: "Leave off your manual labor; give over your bodily toil; bestow your skill and reason to the directing of my power, and I will bear the toil, with no muscle to grow weary, no nerve to relax, no breast to feel faintness." What further improvements may still be made in the use of this astonishing power, it is impossible to know, and it were vain to conjecture.[11]

Power had been about muscle for all of human history, and the most effective way to marshal it had been by assembling serfs and slaves. Now, by golly, the best way was to get yourself a steam engine.

THE NINETEENTH CENTURY is sometimes called the age of coal and steam and ours the age of oil and the internal combustion engine, but the former age continued on into the next century and is still with us. Coal, for instance, is at present the most important fossil fuel for making steam to spin our dynamos and produce our electricity, which the peoples of the richer societies consider as

[10]Ibid.

[11]*The Works of Daniel Webster.* Vol. 1 (Boston: Charles C. Little & James Brown, 1851), 186.

much their birthright as (ironically) air to breathe. Furthermore, we now have techniques to transform coal into liquid combustibles and other essentials, and there are many nations—China, for instance, and, increasingly, the United States—with a good deal more concentrated energy as coal than as oil under their soils. The hard black fuel has continued and will continue as a crucially important factor in world history. The effects of the first stage of the fossil fuel revolution that appalled and entranced the witnesses quoted above fester and foster today in developing nations, from the grim *machiladoras* of Mexico's Tijuana to the silky nightclubs of China's Shanghai.

Coda:
Nellie Bly Beats Phileas Fogg by Eight Days

Phileas Fogg, hero of Jules Verne's trendy novel, *Around the World in Eighty Days* (1873), completed his circumnavigation in eleven weeks and three days. Taking the Suez Canal instead of the Cape of Good Hope route saved him weeks, but his greatest advantage was that he traveled by coal and steam, not sail, i.e., by a much more concentrated product of sun energy than wind. (Incidentally, he ran out of coal and had to burn his vessel's superstructure under the boiler of its steam engine to complete the final leg of his voyage—an obvious echo of what happened to the *Sirius* off New Jersey in 1838.)

A decade and a half after the publication of Verne's book, Nellie Bly, a reporter at the *New York World*, suggested to her editor that the newspaper should sponsor her in an attempt to beat Phileas Fogg's record. He tried to dissuade her, saying that it was not a job for a woman, who would have too much luggage and wouldn't have a gentleman to protect her. She answered that if the editor sent a male reporter around the world, "I'll start on the same day for some other newspaper and beat him."

She won that argument and left to go round the world with one bag and no protector on the morning of November 14, 1889. Six days later she debarked in England. On arriving in France she visited with Jules Verne, and then continued on to Brindisi, Italy, where she caught a ship for Egypt and on to Colombo, Ceylon. She was booked for immediate departure from Colombo for China, but there was a delay of five days. Then she had a piece of luck, a record-setting passage to Hong Kong in spite of headwinds and monsoon storms.

In Hong Kong she saw the American flag at the United States Consulate, the first she had seen since leaving home. "That is the most beautiful flag in the world," she said, "and I am ready to whip anyone who says it isn't." No one spoke. From China she traveled on to "the Land of the Mikado," where she spent 120 hours. Her passage across the Pacific was stormy, but she arrived hale and hearty in San Francisco on January 23, 1890, where she was welcomed enthusiastically. She had completed 21,000 miles in sixty-eight days.

Snows blocked the passes through the Sierra Nevada, so Nellie took the southern route through the Mojave Desert and crossed the continent in four and a half days, reaching Jersey City, her embarkation port, at three-fifteen in the afternoon of January 25.[12] The circulation of the *New York World* was soaring, and she was a media celebrity, one of the very first. There was a popular song entitled "Globe-Trotting Nellie Bly," and soon a Nellie Bly board game that followed square-by-square every day of her circumnavigation. She had circled the world in seventy-two days, six hours, eleven minutes, and fourteen seconds, beating Phileas Fogg's fictional record by more than a week.

One more note about Nellie Bly. She not only proved that you

[12]I have assumed that all the locomotives and ships involved in this journey burned coal, not wood. I do so because wood was in short supply in many regions and these trains and ships were whenever possible chosen for their speed. Some American locomotives were still burning wood circa 1890, but surely not those that carried Bly across North America in four and a half days.

could go round the world in less than eighty days, but also that speed at a rate of a new society a week often promotes nothing so much as fatigue. She summed up what she had learned in her great adventure thus:

> There is really not much for Americans to see in the foreign lands. We've got the best of everything here; we lack in nothing; then when you go over there you must be robbed, you get nothing fit to eat and you see nothing that America cannot improve on wonderfully.[13]

The revolution instigated by Newcomen and Watt only served to confirm what Bly had already known to be true when she climbed the first gangplank.

[13]Brooke Kroeger, *Nellie Bly: Daredevil, Reporter, Feminist* (New York: Random House, 1994), 168–69.

5

OIL AND THE ICE
(Internal Combustion Engine)

For God's sake, be economical with your lamps and candles! Not a gallon you burn, but at least one drop of man's blood was spilled for it.

—Herman Melville (1851)[1]

Oil is probably more important at this moment than anything else. You may have men, munitions, and money, but if you do not have oil, which is today the greatest motive power that you use, all your other advantages would be of comparatively little value.

—Walter Long, House of Commons (October 1917)[2]

A concern of humanity for millennia and its obsession for the past two centuries has been to find ways to tap energy so as to maximize its availability wherever and whenever it is wanted. The pairing of coal and steam to drive engines was an enormous advance in that search, but the search didn't—couldn't—stop there. Humanity always wants more power in smaller units of volume and weight, of greater portability. Humans want to carry sun-power like six-shooters in a cowboy movie. For that they would need new fuels and new prime movers.

By 1900, locomotive engines were operating at five times the

---

[1]Herman Melville, *Moby-Dick; or The Whale* (Harmondsworth, UK: Penguin Books, 1972), 306.
[2]Daniel Yergin, *The Prize: The Epic Quest for Oil, Money, and Power* (New York: Simon & Schuster, 1992), 177.

pressure and efficiency of those that propelled the *Rocket* in 1830. Engineers and mariners had discovered that propellers drove ships through water more efficiently than paddle wheels, and soon there were bigger and faster steamships. But all that was little more than variation on existing technology. A prerequisite for a real breakthrough was a fuel of greater energy density than coal and easier portability. Coal came in chunks and pieces, with a lot of space in between. You could grind it into a powder, but powders even as fine as dust include a lot of air, too, and are miserably unpleasant to deal with. Coal was awkward to move: you couldn't *pour* it, so to speak. At Port Said, Nellie Bly sniffed that her ship was re-coaled by "hurrying, naked people rushing with sacks of coal up a steep gangplank . . . every one of them was yelling something that pleased his own fancy and humor."[3] There must be some more efficient way to distribute domesticated sun energy.

The answer was petroleum, usually called oil in American English. Often a more recent legacy of past sunlight than coal, oil is a product of similar processes. The energy density of oil is roughly 50 percent higher than most coals. It also has the advantage of being a liquid and is therefore more compact, easier to store, and easier to transport than coal. You can run it from point A to point B through pipelines. You can even run it from airplane to airplane through hoses suspended in midair.

Nevertheless, its usefulness had been overlooked for thousands of years. It was used for a miscellany of purposes—weaponry, waterproofing, lubrication, salves, once in a while as a fuel—but there was not a strong demand for the sticky, smelly substance that oozed out of the ground here and there. That demand wouldn't come until we spent a century or so learning that we couldn't depend on whales for the means to lengthen our days and shorten our nights.

[3]Nellie Bly, *Around the World in Seventy-two Days* (New York: Pictorial Weeklies and Brentanos, 1890).

Homo sapiens isn't and never was a nocturnal species. Humans and their hominid ancestors had spent their nights sleeping or at least hiding. Campfires and fireplaces supplied some light, but not much. Torches supplied some more, but were undependable and smoky. Upper Paleolithic artists utilized primitive lamps—probably not much more than wicks floating in dishes of fat—and painted masterpieces in what otherwise would have been pitch-black caves, but that was despite, not because of, the quality of the artificial lighting. Twenty thousand years later, the enormous majority of humans still went to bed not long after sunset because there was not much else you could do in the dark.

The eighteenth century in European history is known as the Age of Enlightenment, which refers to the philosophical innovations of the time and not to an increase in nighttime illumination, but the title works well for both. Both the *philosophes* and streetlights were products of a nascent faith that the quality of human life could be intentionally improved. That would, of course, require a lot more sunlight energy. European cities were growing, along with appreciable numbers of people with enough money to frequent the new coffeehouses to eat, drink, chat, and celebrate after sundown. There were also a lot of urban poor and literally dislocated people from the countryside whose acquaintance the celebrants wouldn't want to make in dark streets.

London will serve nicely as stand-in for all the big cities of Europe and America. It had been long noted as especially dark, probably because of its latitude and the consequent long winter nights. Then in the seventeenth century a few streetlights appeared. By 1700, the City (when capitalized that title refers specifically to the municipal district inside or near the ancient stone walls) shone with streetlights from 6:00 p.m. to midnight for 117 of the darker nights annually. The year 1700 will do as the first marker of our shift from a genetically inherited diurnal cycle to the present self-directed cycle, a shift whose effects—bodily, psychologically, socially—we don't yet fully understand.

Lamplighter refilling an oil lamp in London; anonymous engraving from Knight's *Old England,* 1800.

294.—London Lamplighter, 1800.

By 1736, the City enjoyed 4,000 hours of streetlighting annually, and London ascended to the status of one of Europe's brighter metropoli. The lamplighter, with ladder, fuels, wicks, and all, became a common sight. The Age of Enlightenment was underway, but even so, all of London's streets were dark at night some of the time and most of them all of the time. One evening in 1763 James Boswell, to cite one illustrative example, took full advantage of the favors of a "jolly young damsell" on Westminster Bridge with no concern whatsoever about being seen and interrupted.

Gloom was cheap and light cost money, so illumination inside the houses and shops was also, by our standards, inadequate. Common candles were made of tallow and smoked and smelled and guttered and went out. Lamps didn't give much light and also smoked and smelled and often went out. Their design was poor and you had to keep trimming the wick. Their fuels, vegetable oils and animal fats, weren't consistently flammable and didn't soak upward through the wick as nimbly as a proper illuminant should. Benjamin Franklin reasoned that getting more air to the flame via a hollow wick might improve performance. He tried using a bullrush as a wick, but it was too narrow, and he was soon too busy being a rebel and diplomat to pursue the idea.

Ami Argand, a chemist, friend, and associate of Watt and Boulton, and an innovator who might have been one of the first humans to ride a balloon aloft but for his weak heart, independently came up with the same idea. His lamp wick was held between two concentric metal tubes: the outer tube supported the wick; the inner was hollow, and air rose through it, improving combustion. He also added a glass chimney to protect the flame and increase the draft. In 1783 a contemporary wrote, "The effect of this lamp is exceptionally beautiful. Its extraordinary bright, lively, and almost dazzling light surpasses that of all ordinary lamps. . . ."[4] Argand's lamp produced less smoke and up to ten times more light than previous lamps. When he died in 1803, there were tens of thousands of his lamps illuminating streets, shops, and homes on both sides of the Atlantic. (He should, like Watt, have been rewarded with riches, but the French Revolution, with its chaos and inflation, made sure that he wasn't.)

Argand's lamp inspired a "prodigious importation of whale oil"[5] in Britain, France, and elsewhere. Oil obtained from whale blubber burned more evenly than the other illuminants and with less smoke; sperm whale oil was the very best of all. Such oil meant brighter nights for some and wildly dangerous adventures for others, particularly Americans, who set off on a worldwide pursuit of the sperm whale.

---

[4]Wolfgang Schivelbusch, *Disenchanted Night: The Industrialization of Light in the Nineteenth Century*, trans. Angela Davies (Berkeley: University of California Press, 1988), 11.

[5]John J. Wolfe, *Brandy, Balloons, and Lamps: Ami Argand, 1750–1803* (Carbondale, IL: Southern Illinois Press, 1999), 124.

Sperm oil, according to John Adams, arguing with the British prime minister about admitting American exports soon after the United States gained independence,

> gives the clearest and most beautiful flame of any substance that is known in nature, and we are all surprised that you prefer darkness, and consequent robberies, burglaries, and murders in your streets, to receiving . . . our spermaceti oil. The lamps around Grosvenor Square, I know, and in Downing Street, too, I suppose, are dim by midnight and extinguished by two o'clock; whereas our oil would burn bright till 9 o'clock in the morning, and chase away, before the watchmen, all the villains, and save you the trouble and danger of introducing a new police into the city.[6]

In the last years before the American Revolution, London had been spending £300,000 annually for whale oil for streetlamps. Most of it came from the North American colonies, two thirds from Nantucket, an island off the coast of Massachusetts where the residents had made themselves into the best whalers in the world. The Basques, Dutch, English, Germans, and other Europeans had been whaling in the northern Atlantic for several centuries, but the object of their search in those waters was and would continue to be chiefly right whales and other species whose blubber provided an illuminant inferior to that obtained from the sperm whale. "Les Nantuckois," as Thomas Jefferson, American minister to France, called them, led the way in pursuing sperm whales in warmer waters.

For several generations of whalers, American and European, the perennial problem was not one of demand but of supply. The demand for sperm oil for streetlights, lighthouses, and for lamps in millions of homes slacked off only during wars. The whalers' prob-

---

[6]Quoted in Alexander Starbuck, *History of the American Whale Fishery* (Secaucus, NJ: Castle Books, 1989), 85.

lem was that they were too effective as hunters for their own good. Although limited in power to wind and human muscle, they were able so to thin out whale populations in vast areas of the oceans (such as the Madeira grounds, the Twelve-Forty grounds, the Delagoa Bay grounds, and the Japan grounds) as to render hunting there economically unrewarding. Then they would sail on, discover another whale-rich expanse of ocean, and set to work decimating the giants there. By the 1770s they were off the coasts of Africa and Brazil and as far south as the Falklands. Edmund Burke declared that there was "No sea but that is vexed by their fisheries," and the whalers weren't even around Cape Horn yet.

They accomplished that in the 1790s, rushed into the Pacific, and in a little more than a half century decimated several whale, especially sperm whale, populations in that ocean, the vastest feature on the surface of our planet. In 1848, a whaler from Sag Harbor, New York, 11,000 miles around the Horn from home, passed through the Bering Strait to hunt bowhead whales under the unblinking sun of the Arctic summer. At first, when whales were plentiful, voyages of two years filled a whaling ship's holds with barrels of oil and profits easily exceeded expenses. Now there were voyages of four years and even five years that didn't pay. Humanity's yearnings for the convenience of light at night, though implemented through what we would call a primitive technology (find a fat whale, row up to it, and stick it with a spear), were endangering whole species of *Cetacea*.

There were two choices. One, conservation, unthinkable in the middle of the nineteenth century except by prophets like the primal environmentalist George P. Marsh; or, two, find a substitute for sperm whale oil. That substitute would require new raids on Carboniferous accumulations.

Back in the 1790s, William Murdock, an associate of Watt and Boulton, had discovered how to make coal gas by heating coal in the absence of air, and how to store the gas and distribute it by pipeline. By the middle decades of the nineteenth century, coal gas was displacing whale oil lamps for city lighting, public and private. But it was far from an ideal illuminant. Storing it and piping it to the customer was burdensome and expensive, and it was poisonous and explosive.

Kerosene, which Abraham Gesner, a Canadian chemist, discovered how to distill from petroleum in 1853, proved in many circumstances to be a better choice. It was almost as good an illuminant as sperm oil, and much cheaper. But would there be enough to satisfy a soaring demand for kerosene? In 1859, E. L. Drake, previously a ne'er-do-well, was searching for it at the suitably named Oil Creek in Titusville, Pennsylvania. He rejected the idea of actually digging for it, and chose to seek it out with a drill driven by a small steam engine. He hit it on August 29 at 71 feet, initiating America's and the world's first petroleum rush, much like California's gold rush of ten years before. Before that boom ended in 1879, Oil Creek spouted 56 million gallons of petroleum, kerosene lamps were spreading everywhere, and the American whale fishery was a business of minor importance.[7]

Then the history of human exploitation of sun energy took another violent turn, as it would again and again in the industrial age. The revolution in lighting did not cease with the gaslight and kerosene lamp any more than it had with Argand's lamp. A few decades after the Oil Creek rush, something brighter and safer, Edison's electric light (about which much more will be said in the next chapter), began to displace the preceding means of illumination. That advance would have blighted the demand for oil but for another technological revolution, this one in transportation, which sustained and multiplied humanity's addiction for this kind of fossilized sun energy.

By the middle of the nineteenth century the steam engine was

---

[7]Whaling, revived by new technology—harpoon cannons, steam-powered factory ships, and catcher boats—continued and even expanded, but not in America and not in answer to the demand for light.

providing the means to transport large numbers of people and heavy, bulky freight between major and even minor cities, but only when the demand for such service was strong and prolonged enough to justify the long and expensive effort of building railroads and large steamships. But what about shorter distances and the needs of small groups and individuals? Horses were useful, but tired easily, were sometimes uncooperative, and often couldn't be repaired. The steam-driven automobile, which first appeared early in the nineteenth century, might do as a substitute for the horse; but its engines and boilers were awkward and heavy and it was not to prove broadly adaptive.

In steam engines the burning of the sun-source fuel transformed water into steam, which then drove the pistons. That struck some engineers and inventors as indirect and inefficient. Why not tap the burning fuel directly? Why not in some way or other burn it *inside* the piston? Wouldn't an *internal* combustion engine (ICE) be better than a steam engine? The first experiments along this line were those that had spurred Newcomen and the other early inventors working with the steam engine. In the seventeenth century some visionaries, inspired by firearms, tried to build gunpowder-piston engines. They failed because there was no way to control the rate of explosion or to recharge the cylinder with gunpowder after the explosion, but the idea of an ICE persisted.

Men of many nationalities took up the idea, but it was the Germans who contributed the most to the invention of the basic ICE. In 1863, Nikolaus August Otto, a traveling salesman turned engineer, built one reminiscent of Newcomen's steam engine, as had others in prior decades. The fuel, an illuminating gas (probably coal gas), exploded inside a vertical cylinder, drove the piston up, and then the piston's weight and atmospheric pressure pressed it back down. That engine was a good start and won him some fame, but it wasn't the powerful, smoothly running machine he had in mind.

Theory is fast but application is slow; Otto worked for more than a dozen years trying to make a machine that could safely contain a series of explosions and harness their power. His common sense dictated that the number of piston advances and retreats per power stroke should be minimized. (In the Newcomen engine every other stroke—i.e., the atmosphere stroke—was a power stroke, and in the newer, double-acting engines, with steam pushing the piston to and fro, all strokes were power strokes.) Another sensible admonition forbade sharp compression of the fuel mixture because that might set off unscheduled and dangerous explosions. But common sense doesn't always know what it is talking about.

In 1876, he patented his revolutionary four-cycle engine. The Otto engine, as it is called, functions as follows. Energy supplied by a battery or hand crank or some other outside source initiates the first step of four, drawing the piston back, which sucks fuel and air into the cylinder. Then the piston advances, compressing the fuel-air mixture: this is crucial to assuring full and even combustion. A flame or spark ignites the mixture and a nuanced explosion drives the piston back, the one power stroke of four. In the final stroke the piston advances, expelling the exhaust gases. Now the engine is running on its own and should in theory continue to do so until the fuel is exhausted.

Steam engines needed a long warm-up period and had to be kept hot to be available for use. Otto's engine was not only more efficient per unit of fuel than comparable steam engines, but could be turned on and off as needed. Steam engines, with few exceptions, were big and heavy. His kind of engine was lighter and more adaptable and convenient for shopwork and transportation. Applications of Otto's invention came in a rush. By 1900, there were perhaps 100,000 such engines bearing his name and innumerable others of similar design churning away in machine shops, breweries, pumping stations, and on the road.

There would be other kinds of internal combustion engines, notably the diesel (an invention of another German, Rudolf Diesel), which functions efficiently on a diet of unrefined oil. In time the turbine engine appeared, an ICE that is especially efficient and,

Karl Benz and an assistant seated in the Benz Motorwagon,
the first automobile to be sold to the public, 1886.

with adaptations, especially well suited as the jet engine of faster air-craft. Only the modern rocket engine is more efficient, but it has limited applications. Otto's was the first practical ICE and it is still with us under millions of automobile hoods.

Karl Benz, Gottlieb Daimler, and Wilhelm Maybach, associates of Otto, were protagonists in the next chapter of the story of the ICE. They invented a carburetor that mixed gasoline and air efficiently. (Gasoline, a dangerously flammable distillated crude oil, had previously been considered a worthless byproduct of kerosene production and was usually discarded.) In 1885, they produced not the original automobile but the first that actually worked, a three-wheeler. Five years later, the first Mercedes-Benz automobile rolled out onto the road (and thus immortalized Mercedes Jellinek, the daughter of Emil Jellinek, a wealthy automobile fancier).

The ICE even empowered humans to fly. In 1903 the Wright brothers, Orville and Wilbur, flew, one at a time, in a heavier-than-air vehicle. The most onerous of their problems had been the building of an engine powerful enough to raise their plane up off the ground and speed it along in the air and yet light enough to be carried aloft. Clément Ader, France's nominee as the inventor of the first plane actually to fly, had tried a steam engine. His *Avion III* was a bat-winged craft with a pair of 20-horsepower steam engines that may have gotten off the ground in 1897. But steam engines with boilers filled with water were simply too heavy per horsepower, so the Wright brothers designed an engine, an ICE, of their own. Their 1903 engine produced 12 horsepower and weighed, with fuel and all accessories, only 200 pounds. Their flights, Orville wrote, were "the first in the history of the world in which a machine carrying a man had raised itself by its own power into the air in full flight, had sailed forward without reduction of speed, and had finally landed at a point as high as that from which it started."[8]

The airplane was the most spectacular manifestation of the advantages of ICE, but for most of humanity the earthbound motor vehicle was clearly more important. At first, automobiles were thought of as successors to the carriage, playthings for the wealthy. The Americans, led by Henry Ford, changed that through mass production and mass advertising. In 1903, this son of Irish immigrants and former engineer with the Edison Illuminating Company (see the next chapter) launched the Ford Motor Company with his Model-A. Five years and twenty models later, he presented the world with the first Model-T, the famous "Tin Lizzie," eventually available in nine models ranging from a two-passenger touring car

---

[8]Charles H. Gibbs-Smith, *The Aeroplane: An Historical Survey* (London: Her Majesty's Stationery Office, 1960), 41.

to a light truck, all on the same chassis. The Model-T was strong, durable, and easily operated and repaired by amateurs. After Ford got his assembly lines moving, it was an inexpensive means of individual and family transportation. (His assembly lines were driven by electric motors—more about that in the next chapter, too.) At first his workers took 750 minutes to assemble the major components of a Model-T. In time they got that down to 93 minutes.

In 1925, a Model-T cost $260 and middle-class people—even workers in the Ford plants—could afford one for themselves. Ford Motor Company produced 16 million Model-Ts before discontinuing that model in 1928. Ford's company, of course, continued to produce motor vehicles, now in competition with Packard, Pierce-Arrow, Austin, Morris, Fiat, and others. The auto industry flooded the richer and, increasingly, the poorer nations with automobiles, and the nations bound themselves together with meshes of surfaced roads. (Asphalt, by the way, is, like gasoline and natural gas, a kind of petroleum.) Hundreds of millions of people began to think that driving a vehicle powered by an ICE was a normal, even indispensable, human behavior.

The vehicles burned fossil sun power, gasoline usually, the demand for which soared. The petroleum seekers who had started at Oil Creek field in Pennsylvania rushed on to California, Texas, Oklahoma, and elsewhere in the United States. By 1900, they were pumping oil in Romania, in Baku on the Caspian Sea, and Sumatra, in the Dutch East Indies. By World War I, they had moved on to Mexico, Iran, Trinidad, and Venezuela. In the middle decades of the twentieth century, they were discovering new fields in every continent but Antarctica, most notably in the Middle East, and were even beginning to drill into the sea bottom.

The ICE powering the automobile, truck, and tractor has for a century been the most influential contrivance on the planet. It has allowed us to move from farm to city and helped us at the same time to increase agricultural production immensely. It has allowed us to live in cities without tens of thousands of horses and their torrents of urine and tons of feces. At the end of the twentieth century there were fewer horses but a half a billion cars in the world, not including trucks, buses, tractors, tanks, and so on. Our factories were turning out nearly 50 million new ICE vehicles a year, and we humans required 70 million barrels of oil daily.

During the twentieth century the addiction to oil became a major factor in international affairs and figured in war after war fought by the technologically advanced nations. World War II (1939–45) is a case in point. The leaders of Germany and Japan were convinced that their nations must free themselves of dependency on others for oil. The Soviet Union had oceans of its own oil and France, Britain, and their allies could draw on other oceans of oil under the soils of the United States and elsewhere overseas. Germany and Japan, with no such sources, lusted after the fossil fuel riches of Romania, the USSR, and the Dutch East Indies; they dispatched armies and navies to seize them.

The armed forces of World War II were largely propelled by oil burning in the cylinders of ICEs. The conflict began in Europe with the *Blitzkrieg* (lightning war), a sudden and rapid advance spearheaded by German tanks and planes. It ended with a Soviet assault on Berlin in April 1945, which included 8,000 tanks and 11,000 planes. The ICE enabled the belligerents to wage war not only on enemy armed forces but on their civilian populations as well. The war in the Pacific began, as far as the United States was concerned, with a Japanese air raid on the American military bases in Oahu, Hawaii. It ended with American raids on Japanese cities; on March 10 and 20, 1945, American planes bombed four of these, leveling 83 square kilometers of buildings and killing about 100,000 people. A month later, American B-29s delivered two nuclear bombs, killing at the very least another 100,000. The ICE had enlarged the battlefield to take in every square foot of the Earth's surface.

CODA:
THE ICE, PEACE AND WAR, AND THE TAXIS OF PARIS

During 1914 the ICE enabled France to survive invasion. In August of that year, German armies swept through Belgium into France and swung round and south toward the capital of France. On September 6, they were thirty miles from Paris. What ensued was the First Battle of the Marne, September 6–12, 1914, which saved the city.

One of the key factors in that Allied victory was the four thousand (some sources say six thousand) reinforcements which the French were able to get from Paris to the front on the critical days of September 7 and 8. These soldiers could have marched the thirty miles, but that would have left them exhausted and would have taken at the very least a full day, by the end of which the battle and possibly the war might have been lost. Horses? Impossible. There was no way instantly to procure thousands of horses, much less fodder for them and wagons for them to pull. Steam locomotives could have towed the reinforcements to the front in minutes, but all the railroads within a hundred miles of Paris were jammed with divisions retreating, divisions advancing, with supply trains going no one knew where anymore, with trains loaded with wounded and thousands of refugees. Near Troyes, some trains were taking twenty-four hours to travel six miles.

General Joseph Gallieni, military governor of Paris, ordered the requisition of the hundreds of taxis in Paris. Gendarmes stopped taxis and proclaimed, "Requisition. Go back to your garage."

"But I have a fare," the cabbie would often answer.

"Never mind. Your fare must get out."

"How will I be paid? By meter or flat rate?"

"By the meter."[9]

The taxis gathered at the Esplanade des Invalides and other staging points. There they took on their cargoes of French soldiers, the *poilus*, and careened off to the front, sometimes two and three abreast. The fortunes of war wavered back and forth, and then for the first time in weeks tilted in favor of the Allies. By the twelfth of the month, the Germans were in retreat. They would never come as close to Paris again in that war.

For their services in the First Battle of the Marne, the Parisian cabbies were paid 27 percent of the amounts registered on their fare meters and were elevated into the fellowship of France's legendary heroes.

[9]Georges Blond, *The Marne: The Battle That Saved Paris and the Course of the First World War*, trans. H. Easton Hart (London: Prion Books, 2002), 180.

6

ELECTRICITY

. . . The dynamo itself was but an ingenious channel for conveying somewhere the heat latent in a few tons of poor coal hidden in a dirty engine-house carefully kept out of sight; but to Adams the dynamo became a symbol of infinity. As he grew accustomed to the great gallery of machines, he began to feel the forty-foot dynamos as a moral force, much as the early Christians felt the Cross.

—Henry Adams (1905)[1]

In the middle decades of the nineteenth century humanity continued to lust after more power, but power per se was not the immediate problem. One heat engine might be more powerful than the muscular strength of a thousand or more servants, *but—* a very important but—there is more to getting work done than power. An engine powerful enough to bend steel girders on this side of town is useless if what you want is to fold newspapers on the other side of town. Muscle servants, especially the human ones, can be easily moved from this side of town to where their strength is needed. They can be switched from job to job, employing their muscles for hod-carrying *here* in the morning and for shelling walnuts over *there* in the afternoon. Human muscles can be utilized for tasks as crude as hacking trails through the jungle and as delicate as painting scenes on porcelain teacups.

[1] *Henry Adams: Novels, Mont Saint Michel, The Education* (New York: Literary Classics of the United States, 1984), 1067.

Heat engines tend to be immobile, especially the giant ones that produce enough power to drive whole factories. The steam locomotive and the gasoline-driven automobile certainly move, but characteristically only along prepared roads. A heat engine might lift 100 tons, but was, in itself, not useful if what you wanted to do was drive one sewing machine in a millinery shop up two flights of stairs and down the end of the hall or light a lamp on a farm a hundred miles out in the backcountry. Distribution, focus, differentiation, and gradation of prime mover power were the challenges for a society that wanted more material goods, more convenient transportation, more light—more of *more*—in the middle decades of the nineteenth century. The answer was electricity.

Evidence of electricity had, of course, always been present in our lives, from lightning bolts blasting steeples to little boys scuffing along a rug in cold weather to deliver shocks to little girls' noses. The ancient Greeks discovered that rubbing amber somehow changed it so that it attracted feathers (the word "electricity" is derived from the Greek name for amber). But this added nothing to the possibilities of humanity ever being able to do anything useful with the phenomenon. Very little would happen in that category until ways were found to evoke electricity in forces great enough and of sufficient duration to be a subject for thoughtful examination.

That started to happen when William Gilbert, Queen Elizabeth I's physician, learned that there were a number of substances in addition to amber that gained magnetic qualities when rubbed. A hundred years later another Englishman, Francis Hawksbee, created a real electrostatic machine by attaching a crank to a hollow glass globe, rotating the globe at high speeds, and rubbing it with a leather pad. This machine produced sparks for entertainment and enough electricity for real experiments.

Now some way to store electricity for deliberate and thoughtful consideration was needed. The initial answer to that need was the Leyden jar, named after the city in Holland where Pieter van Musschenbroek is supposed to have invented it in 1746. The first model was simply a jar of water with a wire sitting in it vertically, half in, half out of the water. In later models the jar was wrapped in metal coating, inside and out. The wire was charged with electricity from a big version of Hawksbee's machine. The charge, caught in the bottle, persisted for some time, even for several days, to be summoned when wanted.

Electrostatic machines and Leyden jars provided electricity in jolts of energy suitable for tricks and some experiments, but not really suitable for the kind of investigations that would lead to major advances in scientific understanding. What scientists needed was electricity in a steady flow. An Italian, Alessandro Volta, provided that toward the end of the eighteenth century amidst the confusion and fury of the French Revolution. The Leyden jar stored electricity. Volta's invention made electricity.

He did so with chemistry. He stacked copper, zinc, and cardboard disks moistened with saltwater, building what we call the voltaic pile. The copper loses electrons to the wet cardboard and the zinc gains electrons. Unattached electrons flow out of the pile through insulated wires. This, the first battery, produces electricity until the fluid dries out or the zinc dissolves. It was a marvel and Napoleon took time off from battles to give Volta a gold medal, the cross of the Legion of Honor, and 6,000 francs.

Within a few years there were piles weighing tons—the Royal Institution in London had one with 4,000 copper and zinc disks—and other and better batteries soon appeared to provide science and technology the means to investigate and perhaps to domesticate a thoroughly mysterious force of nature.

In 1820 a Danish professor, Hans Christian Ørsted, noticed in the midst of one of his lectures on electricity that the wire he held in his hand, connected to a small Voltaic battery, was somehow affecting the needle of a compass on the desk. He subsequently discovered that no matter where north was, the needle would position itself at right angles to the wire. If he reversed the flow of electricity through the wire, the needle swung 180 degrees and took up the opposite position. Electromagnetism can be tapped to make things move; it can do work. Electric motors are descendants of his tiny experiment.

The next electromagnetic hero was an Englishman, Michael Faraday, whose career provides the best kind of evidence in favor of meritocracy. He was one of a blacksmith's ten children and a man of very little formal education, who happened to attend a public lecture by the famous Sir Humphry Davy. He came away fascinated, wrote to the scientist, became his assistant, and in time succeeded him as head of the Royal Institution laboratory. Faraday repeated and elaborated on Ørsted's experiment, and soon had not only needles rotating but disks revolving. That was satisfying; but, he wondered, if electricity and magnetism could be manipulated to produce movement, would it work the other way around? In 1822 he wrote in his notebook: "Convert magnetism into electricity."

Nine years later, Faraday discovered that when he thrust a bar magnet in and out of a coil of wire attached to a galvanometer, the instrument indicated that a current of electricity had been created. Magnetism *and motion* made electricity. Faraday proceeded to build the first dynamo or generator, a metal disk rotated by a hand and crank between the poles of a horseshoe magnet, producing a weak flow of electricity. (Joseph Henry made much the same discovery in America at approximately the same time, but Faraday published first.)

Ørsted and Faraday had discovered an intimate relationship between electricity, magnetism, and motion. Given control of the first two, humans could produce motion. Given control of the second two, humans might be able to produce electricity. The first descendants of Faraday's dynamo were quite similar in design—a loop of wire spinning between the poles of permanent magnets, or vice versa. The volume of electricity that was produced was insufficient to drive large motors, but it was enough to inspire inventions that opened the era of worldwide electrification. The United States, just beginning industrialization and overflowing with clever and ambitious young men, would play a central role in the saga of turning European science into fame and dollars.

The first breakthrough invention was the telegraph; it didn't require a lot of electricity to carry a message. Samuel F. B. Morse was an artist who worked in both America and Europe. When he started painting, the best long-range signal system in existence consisted of big semaphores on towers, which was useless in bad weather, tricky at night, and capable of transmitting only the simplest messages. Painting didn't satisfy him, and in time he turned away from his canvases to join the ranks of those who theorized that electric current, which could flow through wires for remarkably long distances day and night, might be suitable for communication. Morse, assisted ably by Alfred Vail, made some useful improvements in the equipment of telegraphy. But his greatest contribution was probably an "alphabet" of dots and dashes by which the staccato opening and closing of a circuit could be utilized to represent letters and numbers. That is what we call Morse Code.

He tinkered with his telegraph for years and spent more time than that pleading for money from governments and private investors on both sides of the Atlantic. (One of his voyages across that ocean was on the swift steamer, the *Great Western*, mentioned in chapter 6—there was a lot going on in the 1830s.) Success came to Morse in the 1840s, about a dozen or so years after Faraday had first published his discoveries on electromagnetism. The climactic moment came on May 24, 1844, when publicity-wise Morse sent the famous and pious message, "What hath God wrought?" down a wire that stretched from the Mount Clare depot in Baltimore to the Supreme Court Building in Washington, D.C. A dozen years later there was a telegraph line connecting Washington and New York.

The Morse telegraph system answered a powerful need and spread with amazing speed. In 1861 the first transcontinental telegraph signal pulsed along a line on poles stretching from San Francisco to New York. In 1857 and 1858 the first attempts at laying a telegraph cable between continents, specifically from Newfoundland to Ireland, ended in failure, frustrating but not stopping the pioneers of electricity and of venture capitalism. After a pause for the American Civil War, they renewed the attempt, failed again in 1865, and then in 1866 completed the first transoceanic cable for transmitting messages by electrical pulse. The signal was feeble, but the miracle had been achieved. Now people could correspond instantly across an ocean. The world had for many purposes—economic, diplomatic, intellectual, literary, journalistic—abruptly shrunk by the measure of the width of the Atlantic.

Two decades more and another American, Alexander Graham Bell, made his contribution. (His indispensable aide was Thomas A. Watson, the man who answered the very first telephone call, the impromptu "Watson, come here. I want you," delivered by Bell when he spilled acid on himself.) Bell parlayed his understanding of the processes of human speech production (he had spent years working to provide the deaf with the means to communicate) with what he taught himself and picked up from people like Joseph Henry about electricity to produce the first practical telephone. He patented his invention in the United States in 1876. You could now actually *talk* to someone across town, and soon, with someone in the next city, and soon, other countries. The world's girth diminished again.

Access to electricity revolutionized communication, but would do so across the whole range of human activity—city lighting, industry, transportation, entertainment—only if massive voltages became available. Faraday's dynamo had produced just a wisp of current. Its early successors, which, like the original, used natural magnets, produced more electricity, but not enough more to power a new civilization. Natural magnets were simply not powerful enough.

Joseph Henry and other innovators wrapped current-carrying, insulated wires around big horseshoes of soft-core iron and turned them into electromagnets, which for all practical purposes could be made as powerful as needed. One of Henry's could lift a ton. In 1866, a German, Werner von Siemens, tried electromagnets in the generator he was attempting to invent, and the electric current flowed like the Rhine. Other Europeans, for instance the Belgian Zénobe Théophile Gramme, made other alterations to dynamo design, improving function.

The prime mover of electricity had been muscle at first (Faraday's muscle spinning the loop between the arms of a magnet), but in the later decades of the nineteenth century generators were being driven by internal combustion engines, at first reciprocating and then, in the final years of the century, turbine engines. The ancient prime mover, the watermill, was adapted for the same purpose, and then the windmill.

With the refinement of electromagnets and dynamos there was enough dependable electricity to drive big electric motors empowering advances in transportation. The first public electric railroad was probably the one that began operation in Lichterfelde, Germany, in 1881. The first London Underground railway (subway in American English) was dug in 1887–90—steam engines were unsuitable underground, but electric motors made no smoke and gave light, too.

The most spectacular advance in electrification was in urban illumination. Back in 1850, there was not a city on earth in which every street was lighted every night and few in which most streets were lighted. The houses of the wealthy were illuminated by candles and lamps burning whale oil, solidified and liquid. Kerosene lamps wouldn't appear in great numbers until the petroleum age opened in a decade or so. Even after that, the best answer for years to the demand for urban lighting was coal gas illumination, which many cities adopted for outdoor illumination and which individuals introduced into their shops and homes.

For most of the nineteenth century coal gas was, on balance, the superior source of light. It burned brightly and could be transported by pipe. Soon many of the larger cities in the Western world had big storage tanks of coal gas called gasometers (which confusingly also referred to meters for measuring amounts of gas) and underground pipes leading to customers' homes, offices, and workshops. Coal gas was much cheaper than whale oil and was widely adopted for street-lighting. London was the first of the big cities to do so. In 1822, it already had four gas companies and forty-seven gasometers.

Gaslight was a great improvement on what had come before, but soon innovators were wondering whether electricity, which experimenters and charlatans far back in the eighteenth century had used to make bright sparks and eerie glows, mightn't prove better. The heat of the flame of coal gas was unpleasant in sultry weather, and in winter, when you shut the windows, it burned up the oxygen in the room and left you with a headache. More than a few textile mills illuminated by gaslight caught fire and burned down. Wind could blow out a gas flame but the gas would continue to flow, setting the scene for disaster. You wouldn't want to leave children alone with the gas's open flame, and the gasometers might explode, wiping out whole neighborhoods.

Such objections were heard more and more often as electric lights became available. The earliest to be used extensively was the arc light. Run a strong current through two sticks of carbon in contact, then draw the two apart and the electricity will arc the gap, burning the carbon and producing a dazzling light. The arc light was (and is) an excellent means to illuminate large areas—city squares and stadiums, for instance—and was utilized to do so for some of the more famous streets of the world and for a few entire neighborhoods. There were plans to light whole cities: imagine the Eiffel Tower decked out with arc lights. But the shadows the big arcs on towers threw on one side of a city street were as black as the other side was bright, and the light was much too dazzling for home use. Imagine a reading lamp with a glaring arc light fizzing and sizzling by your cringing ear.

Dozens of people tried their hand at creating incandescent lamps, piling up a good deal of data from which later inventors benefited. The winner of the race, as the reader knows, no doubt, was Thomas Edison. He was very bright, an extremely hard worker with a fine staff of experimenters, and he hired very good lawyers to defend his inventions and attack his rivals' claims in the courts. His problem was to find some substance—a wick or filament—that would heat up when electricity flowed through it and glow brightly for a good long time. He and his staff tried hundreds of candidates and finally created a lamp consisting of an airless glass bulb containing a filament of carbonized thread. They ran electricity through the filament and it shone for forty-five hours before going out. Edison patented it in 1880. (He and others kept on looking for better filaments; Edison settled on carbonized bamboo for some years, but tungsten became standard in the next century. An inert gas was substituted later for the vacuum.)

Edison's light bulbs did not dazzle and glare like arc lights, provided more light than coal gas, and did not impose a constant threat to ignite wall hangings, clothing, or human hair. They did not consume oxygen or smell in the slightest. (In 1913, Marcel Proust wrote of the "stifling sensation that, nowadays, people who have been used for half a lifetime to electric light derive from a smoking lamp or a candle that needs to be snuffed.") Light bulb energy came by wires, which, unlike gas pipes, could be arranged and rearranged as wanted. You didn't literally have to set fire to an electric bulb as you did a gaslight. Electricity, unlike gas, did not explode. It might electrocute you, of course, but that affected individuals, not whole buildings and neighborhoods.

The most advantageous characteristic of electricity was that it was eminently transportable. You could build your generator by a coal mine, for instance, and burn the fuel right there to produce electricity; or build it next to Niagara Falls and exploit falling water to produce electricity and then dispatch the product through wires (renamed cables when they got big enough) to distant customers. Fossil fuels, including coal gas, had to be ordered beforehand and carried by rail and wagon or run through pipes, which was awkward and consumed time and money. Flip a switch and electricity was right there on the spot—the perfect servant—to drive your locomotive, printing press, elevator, or sewing machine.

Buildings of the Chicago World's Fair illuminated at night;
photograph by C. D. Arnold, 1893.

In 1882, Edison opened the first electric power station supplying private customers at 255–257 Pearl Street in New York City. At first it served only eighty-five customers and lighted four hundred light bulbs, but coal gas was on its way to demotion from being a source of both light and heat to being a source for the latter only. At some moment in the later years of the nineteenth century, Western civilization plumped for electrification. It was an epochal decision soon taken up, at least in intention, by the rest of humanity. It has no precise date: there was no formal Declaration of Electrification. It was not made by a commission or a legislature, but by whole populations informally and gradually over a period of years that began, shall we say, with Morse's telegraph. The best we can do for a date is to find in the record a substantial block of evidence unambiguously demonstrating that it had been made.

The four hundredth anniversary of Christopher Columbus's discovery of America, venerated by citizens of the United States, arrived a half century after Morse's triumph. A commemoration proportionate to the Columbian triumph would have to be truly impressive; a world's fair might suit. Several cities battled for the honor, and Chicago won. That city's hope was to literally outshine Paris, whose World's Fair of 1889 had wowed the customers with Gustave Eiffel's new tower.

The Chicago World's Fair, better known as the Columbian Exposition or the White City, was officially opened at a locale a few miles from downtown on May 1, 1893. (A year late, you may notice, but major demonstrations of a civilization's prowess aren't thrown together in a day.) President Grover Cleveland spoke to a crowd of scores of thousands from a platform seating three thousand:

> As by a touch the machinery that gives life to this vast Exposition is now set in motion, so in the same instant let our hopes and aspirations awaken forces which in all time to come shall influence the welfare, the dignity, and the freedom of mankind."[2]

He pressed down on a key closing a circuit and, as a choir sang Handel's Hallelujah Chorus, an engine burning oil (coal would be too smoky) started running, secondary machines began to function, the fountains to spout, and the biggest world's fair to date officially opened its gates to the public.

The Columbian Exposition was the full-bore affirmation of the supremely self-confident Western civilization expressed specifically by Chicago, which would soon be appropriately called "the city of the big shoulders." The draining and leveling of the site, 686 acres of wetlands, and the raising of its buildings, had been a largely elec-

---

[2]Jill Jonnes, *Empires of Light: Edison, Tesla, Westinghouse, and the Race to Electrify the World* (New York: Random House, 2003), 263.

trified effort. Dredges, stone crushers, sharpeners, the bigger saws, the pumps, the cranes, the spray painters were powered by electricity. The night crew, of course, operated under electric lights. The White City's biggest buildings, all designed by American architects in the classic style with pillars and domes, were grouped around a gigantic lagoon, 250 by 2,500 feet. The biggest single structure, the Manufactures and Liberal Arts Building, was the largest roofed building ever built. It was big enough to have enclosed at one and the same time the cubic footage of the U.S. Capitol, the Great Pyramid, Winchester Cathedral, Madison Square Garden, and St. Paul's Cathedral.

The exposition's prime movers were fifteen steam engines with the aggregate capacity of 13,000 horsepower driving, when required, sixteen generators to produce 8,955 kilowatts of power, enough to light 172,000 incandescent bulbs. The generators supplied the power for all sorts of electric tools, for fountains spouting 100 feet high, for a moving sidewalk as long as the lagoon, a pickpocket detector, an electric chair, and much, much more.

There was the largest searchlight in the world, an arc light that weighed 6,000 pounds. There was an incandescent bulb 8 feet tall on top of an 80-foot column. There was an 11-ton cheese and a 1,500-pound chocolate Venus. The latter weren't directly dependent on electricity, but the spectators wanting to see them arrived on trolleys and boats with electrical motors. There was a gigantic vertical merry-go-round, the creation of George W. G. Ferris, whose name has come down to us as the generic term for such devices. This particular monster of delight weighed 1,200 tons, was 250 feet in diameter, and had thirty-six cars carrying forty passengers each. When full, it carried 1,440 passengers. It was favorably compared to the Eiffel Tower. No wonder the author Hamlin Garland wrote his father, "Sell the cook stove if necessary and come. You *must* see the fair."[3]

[3]Quoted in David F. Burg, *Chicago's White City of 1893* (Lexington: University of Kentucky Press, 1976), 180.

"Nearly everything," David F. Burg, informs us, "that glowed, that sounded, that moved at the fair was powered by electricity."[4] The exposition was open for only half a year, from May to October of 1893, but in that time over 27 million people, Americans and foreign visitors, viewed its marvels. They were awed: electricity seemingly enabled humanity to project and direct the power of its prime movers to any target and for any purpose.

HENRY ADAMS, one of the most perceptive students of human history, concerned himself not so much with dates and details as with the driving forces. He sifted through contemporary civilization for something that might represent, symbolize, the force of his time, and found it in the 40-foot electrical dynamo exhibited at the Paris Exhibition of 1900. He found himself disposed to pray to it, prayer being "the natural expression of man before a silent and infinite force."[5] Many since have been similarly moved. In 1920, Vladimir Ilich Lenin proclaimed like an Old Testament prophet to the All-Russia Congress of Soviets, "Communism is Soviet power plus electrification of the whole country," and the USSR, as soon as it even started to recover from its civil war, began to build electric power plants and some of the biggest hydroelectric dams yet. The United States manifested equal faith in technology by building similarly huge hydroelectric dams—such as the Hoover Dam on the Colorado River, the Grand Coulee on the Columbia, and several in the Tennessee Valley—whose electric energy would prove to be crucially important in the development of the nuclear bombs that ended World War II. Since the war the developing world has followed suite with, for instance, a dam on the Paraná between Paraguay and Argentina, and the epochal Three Gorges Dam now being built on the Yangtze River in China. Per capita consumption of electricity in this world of ours has risen from 1,445 kilowatts in

[4]Ibid., 204
[5]*Henry Adams*, 361.

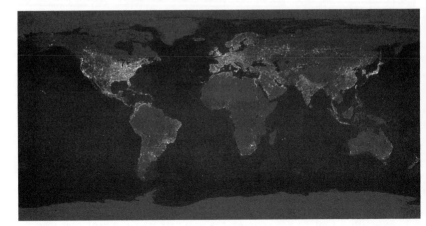

Nighttime Map of Our Planet. A composite of hundreds of pictures taken by orbiting satellites by the National Aeronautics and Space Administration.

1980 to 2,175 in 2000, and will continue to soar in the developing nations.

What is the magnitude of this change? I refer you to the nighttime depiction of our planet above, a composite of hundreds of pictures by orbiting satellites (obviously, it cannot be dark all over the earth at once). Note that the only large areas without visible electric lights are the oceans, polar regions, deserts, and the sparsely settled interiors of the continents—and North Korea. The enormous majority of human beings live within a few feet or, at most, a short walk, of electric outlets.

Coda: Amusement and Execution by Electricity

In the eighteenth century, experimenters learned that human flesh and bone could transmit electricity, which could be amusing. Jean-Antoine Nollet (often known simply as abbé Nol let) contributed significantly to advances in electrical science, but is commonly remembered best for his tricks. For the edification of Louis XV, king of France, he arranged 180 gendarmes in a circle holding hands and had one of them touch the brass ball in the lid of a charged Leyden jar. The shock ran through all 180 instantly and they jumped and gasped in perfect unison. The king loved it. The abbé tried the trick again with two hundred Carthusian monks and yet again with six hundred assorted people at the Collège de Navarre. It was a real crowd-pleaser.

Electric current could be dangerous as well as amusing. In 1750, when Benjamin Franklin was preparing to electrocute a turkey (to eat, probably), he touched what he shouldn't have touched and knocked himself out. The shock, he recorded, was "a universal blow throughout my whole body from head to foot, which seemed within as well as without," but he recovered in a day.[6] The means to create an electrical charge and to store it in the middle years of Franklin's century were such that the electrocution of an adult human being was unlikely. The voltage wasn't high enough. Lightning, however, could provide enough: Franklin, famously flying a kite in a thunderstorm in 1752, elicited a spark when he could have called down a thunderbolt. The next year Georg Richman, a Swedish experimenter, tried a similar experiment in St. Petersburg and was killed.

In the 1880s, dynamos still couldn't match the voltage of thunderbolts, but were quite capable of producing enough to kill a human, which, in fact, had already happened accidentally. It was inevitable that someday it would happen purposely. On March 29, 1889, William Kemmler (alias John Hort) of Buffalo, New York, killed Tillie, his consort, with an ax. He was easily apprehended and convicted—he even seemed anxious for punishment—and the judge condemned him to die by electricity. Perhaps the novelty of such an execution persuaded the judge or perhaps electrocution seemed more merciful than the rope. Kemmler would be the first in the world to be so executed as a formal act of justice.

[6]Quoted in Jonnes, *Empires of Light*, 27.

The sentence triggered a debate in the courts and press about whether such an execution would or would not count as "cruel and unusual punishment," forbidden in the U.S. Constitution. There was also some question as to whether electricity could be depended upon to do the job.[7] Edison confidently testified that electrocution would work, recommending that the subject's hands should be placed in a jar of water diluted with caustic potash and connected to the electrodes. Kemmler himself decided that it would be less painful to die by electricity than by hanging.

The execution took place on August 6, 1890, in Auburn, New York. When the switch on the line connecting a prison dynamo and the electric chair was thrown, Kemmler snapped up straight and strained against his bonds. According to the *New York Times* reporter, he became "as rigid as though cast in bronze, save for the index finger of the right hand, which closed up so tightly that the nail penetrated the flesh on the first joint and the blood trickled out on the arm of the chair."

The current was turned off; doctors examined Kemmler and pronounced him dead. One of the witnesses, Dr. Alfred W. Southwick, an advocate of electrocution as a merciful means of execution, declared: "There is the culmination of ten years' work and study. We live in a higher civilization from this day."[8]

But Kemmler's chest still rose and fell: he seemed to be breathing. The switch was thrown again. Kemmler went rigid again. For an instant there was a blue flame at his neck. His clothes caught fire. There was a very strong smell of burned meat. Now there was no question as to his death.

[7]This debate was one episode in the struggle between the advocates of direct current (DC) versus those of alternate current (AC), which we don't have to go into here.

[8]Jonnes, *Empires of Light*, 211–12.

PART THREE

ENERGY AT THE TURN OF THE THIRD MILLENNIUM

The most fundamental attribute of modern society is simply this: ours is a high-energy civilization based largely on combustion of fossil fuels.

—Vaclav Smil, geographer (2003)[1]

P olitical and military titans such as Caesar Augustus and Genghis Khan had no more than continental effect at most, and usually a great deal less than that. But the changes initiated by the pioneers of modern technology—Newcomen, Watt, and their steam engines, Otto and the internal combustion engine, Edison and electric lights—have been so massive that they would surely attract the attention of any intelligent organisms in passing spaceships. Newcomen and the rest of them utilized new prime movers and new ways to use them to do work and make profits. In doing so, they altered human society and the biosphere drastically and certainly more abruptly than anything since the advent of agriculture and perhaps even the domestication of fire. Estimates of the gross world product (GWP) are incurably imprecise, but they are useful in conveying the magnitude of change. In 1900 the GWP, measured in 1990 U.S. dollars, amounted to something like $1 trillion; in 1950, $5 trillion; and by the end of the century, $29 trillion.

[1]Vaclav Smil, *Energy at the Crossroads: Global Perspectives and Uncertainties* (Cambridge, MA: MIT Press, 2003), 1.

That awesome increase was made possible by mass transportation, giant freighters, elevators, refrigerators, cranes, and other large technologies, and by dynamos producing gigawatts (billions of watts) of electrical power. Almost all of these innovative devices were driven by the burning of fossil fuels or by motors whose electrical energy was derived from combustion of fossil fuels. Most of our electrical energy, despite the photographs of gigantic dams with thousands of tons of water pouring through penstocks as big as subway tunnels and turbines spinning like tornados, is produced by dynamos driven by burning fossil fuels.

John P. Holden, of Harvard's Kennedy School of Government estimates that humanity's primary energy use has increased twenty times over since 1850 and nearly five times over since 1950. The United States, bellwether of the profligate societies, consumed 2,000 kilowatt-hours of electrical energy per individual in 1950, and 32,700 per individual in 2000. Electric plants typically burn tons and tons of coal every day. In 1900, coal production was about 1 billion metric tons; in 1950, 3.5 billion tons; and toward the end of the century, 5.2 billion. Although coal production has continued to rise, oil now supplies more of the world's energy than coal. In 1900, the world production of oil amounted to about 100 million barrels a year. At the end of the twentieth century, the annual total was far over 20 billion barrels.

There is enough coal still in the ground to last us for generations. Our immediate fossil fuel problem pertains to oil, which is the highest in energy density of the fossil fuels and, being a liquid, the easiest to transport and store. The question is not *if* but *when* we will have an oil shortage. It may be generations from now or, as some shrewd analysts predict, in this generation. Kenneth S. Deffeyes, professor emeritus of oil geology at Princeton University, has meticulously considered all the pertinent information, subjected it to sophisticated mathematical analysis, and predicts that world oil production will peak in the first decade of the twenty-first century. After oil production maximizes, whether it be in this decade or decades hence, the trend in oil prices will then be upward, with, one might predict, grim results economically, and politically and militarily as well.

Most, perhaps all, of the billion-barrel, "elephant" oil fields have been discovered and are producing already. Smaller fields will certainly be found, abandoned wells will be squeezed again by means of steam injection and such, liquefying and pushing oil to the surface. Exploratory drilling in waters kilometers deep in the Gulf of Mexico and off West Africa is already underway. But it seems that the gravy days of the oil industry are long gone. We have pumped out and burned in a century the easily accessible accumulations of liquid fossilized sunlight. Fortunately, at least in the short run, there are other sources of oil. Enormous deposits of solid petroleum exist in Canada and Venezuela in tar sands and oil shale to which we may have to resort someday. At present, however, the processes to produce from such sources fuels that are conveniently transportable and burnable are difficult and dauntingly expensive.

Natural gas, which refineries used to burn (or "flare") to be rid of it because it was as worthless as gasoline before the automobile, is now used as fuel for all kinds of purposes. The globe's crust contains huge amounts of it. There is also an icelike form of water (gas hydrates) with cavities containing combustible gas in vast quantities beneath the ocean floor. It is likely that natural gas will take the place of oil as our most important fossil fuel.

The problem is not that we will run out of oil or natural gas in any absolute sense; there will always be some somewhere, most likely in the Middle East, which makes for diplomatic and military complications. But for now let us focus on a relatively simple economic problem: the amount of energy expended to procure petroleum energy is creeping up on the amount of energy gained. For instance, horizontal drilling, one of the clever techniques for getting more out of waning wells, is currently producing in Oman a liquid that is 90 percent water and only 10 percent oil. Getting rid of the water is expensive in energy and therefore in money. It is a matter of "energy returned on energy invested" (EROI)."[2] In the oil fields of the United States the EROI in the 1930s was at least 100 to 1. Today, it is down to about 17 to 1.

The EROI concept was devised by economists, who measure much of what is important but not all. Ecologists examine the *full*—not just the financial—cost of fossil fuels. What happens to our environment when we burn even natural gas, which is relatively clean? We get energy, yes, but also effluvia to dirty our laundry (we could live with that); pollutants to kill our forests; fumes and particulates, including carcinogens, to choke our lungs; and carbon dioxide and other gases to transform the atmosphere into a lid that hampers normal Sun-to-Earth respiration.

Heat engines and electric power plants burning fossil fuels emit millions of tons of sulfur dioxides, nitrogen oxides, and other pollutants into the air. There the sunlight transforms them into other, often more dangerous effluents, which descend to the Earth as acid rain, corroding statues and buildings, leaching soil fertility, poisoning plants, poisoning lakes and rivers, and killing the fish. Acid rain is common downwind of cities and power plants everywhere.

Air pollution from fossil fuels taxes the lungs, causing asthma, bronchitis, emphysema, and other respiratory disorders, including cancer. In Los Angeles the "smog,"—a complicated end product of automobile exhaust fumes drifting aloft in the Southern California sunlight—started people coughing in the 1940s. Citizens of England's capital city had been accustomed to coughing for centuries; then, in December 1952, an especially thick and long-lasting "London fog"—coal smoke—killed over four thousand people. Atmospheric pollution in the cities of the developing world often exceeds that of the cities of the developed world.

The balance of entering and exiting sun energy in our atmosphere is delicate. If too much bounces back into space, the glaciers

[2]Charles Hall et al., "Hydrocarbons and the Evolution of Human Culture," *Nature*, 426 (Nov. 20, 2003), 320.

advance; but if too little exits, global warming ensues, humanity swelters, climates change erratically, and malaria and other tropical diseases spread. This, the notorious "greenhouse effect," has often occurred in the past when volcanoes spewed their gases in abundance into the atmosphere. It is happening now as we burn immense quantities of fossil fuels, releasing into the air in a single year the carbon accumulations of many, many years.

We have to adjust to the long-range realities of our energy situation. One, we can substitute for fossil fuels environmentally friendly sources of energy, which, if utilized wisely, won't pollute and cannot be exhausted. Two, we can find a new, more powerful, and yet clean prime mover.

Let us consider environmentally friendly sources of energy, which include manifestations of sunlight that we have already touched on, starting with biomass. When you burn biomass, you release pollutants, most importantly CO_2, but then, at least in theory, plants grow to replace those burned and absorb the pollutant.

Wood has been our favorite biomass fuel for many thousands of years. Indeed, wood was the most important fuel in the United States until the last years of the nineteenth century. Millions of hectares of the forests of the temperate zone were mowed down to satisfy the needs of modernizing societies for open farm- and ranchland, construction materials, and fuel. The professional logger is as suitable a symbol of the early industrial revolution as the mill worker, and vast clear-cut hinterlands are as characteristic of that period as Dickensian slums.

With the shift from wood to coal and oil, the trees have returned in much of the temperate zone where there is room for them. Southern New England, for instance, has more trees today than in Henry Thoreau's time. If we properly balance harvesting and new growth, temperate zone dwellers might permanently enjoy a supply of this biomass. At present 40 to 50 percent of humans, most of them living in the tropics, still rely on wood for fuel. There the forests, opened up by trucks and chain saws burning gasoline, are being consumed locally for fuel and for massive exports to the rich nations like Japan and the United States. But in the tropics, where the sun can bake the dirt stone-hard and the rains tear it away and carry it off, the trees have not rushed back to reoccupy former forestland.

Wood is available for many purposes, but not in sufficient quantity as the fuel for a newly massive infusion of energy. Even heavily forested Austria, Sweden, and Finland supply no more than 15 percent of their energy needs with wood. Plantations of fast-growing trees like sycamore, poplar, and silver maple are a help, but the energy demand of the industrialized societies of the temperate zone is far too great to be answered with wood of any origin. The situation in the tropics where the forests are shrinking rapidly is worse.

Organic wastes of other kinds—stubble from harvested fields, pulp from paper mills, refuse from sugar cane mills, trash and garbage collected in our cities—are available in quantity. In many cases burning them to produce electricity is cheaper than disposing of them otherwise. Fast-growing plants like switchgrass cultivated for fuel would be of some help, as would tree plantations, but the unpleasant truth is that leaves, stems, and organic debris don't contain enough energy to run our cities for more than a few days per year.

The exploitation of plants such as sugar cane and maize as the raw material for chemical transformation into such fuels as ethanol and methanol is promising. We are already cultivating these and could easily raise more, but we have to be tough-minded about easy breakthroughs. According to some experts, the United States would have to plant maize in most of its entire expanse to produce enough ethanol to replace the gasoline it consumes annually. There is less hope that Europe and Japan, with more people per square mile than the United States, could raise enough in energy crops to fill their needs. And if one asks the question how much energy one gets out of such crops as compared to how much energy one expends to cultivate and process them, the answer is not encouraging. At the present time Americans expend 70 percent more energy in producing a gallon of ethanol than they get out of that gallon. This might be accomplished more efficiently, but probably not 70 percent more efficiently. Ethanol has a truly dreadful EROI.

We are still all children of the sun, and tapping that source directly could solve all of our energy problems. Techniques and devices to do this already exist. Architects know how to maximize passive benefit from the sun by orienting buildings and positioning windows. Engineers know how to trap the sun's heat in various materials, most successfully in water and salt solutions. They know how to focus sunlight with reflectors to heat water to steam to drive turbines. They know how to change sunlight into electricity with photovoltaic cells. These cells are expensive per watt, but have been used with great success in orbiting satellites and in spacecraft and remote locations on earth.

But draping our countrysides with solar cells would fully answer power requirements in only a few sun-baked nations, such as Saudi Arabia. They wouldn't help much in Scandinavia, Canada, and Russia during their dark winters or, for that matter, anywhere at all at night. We can and have come up with practical methods of storing solar energy for delayed use, but that is a complicated and expensive pursuit.

Hydroelectric power has been an essential source of energy for a long time and will continue as such, but most of the rivers suitable for big dams in the developed nations have already been dammed or are filled to the banks with freighters, ferries, and barges. There are suitable rivers for such structures in the developing nations—for instance, a dam at the Three Gorges of the Upper Yangtze will soon be producing an abundance of megawatts of power—but such locations are rare (and rife with social and ecological effects, but that is another story). The future of hydroelectric power probably lies in thousands of unpretentious dams on minor rivers, which will require attention to the difficulties—technological and political—in the collection and distribution of energies gained in small quantities.

Geothermal energy—heat drawn out of the ground—is sub-

tracted from the planet's sum of energy forever, but that total is
practically infinite, at least for the next billion years. Geothermal
energy is available everywhere if you drill deeply enough, but thus
far is tapped for energy in only a few places where magma lifts close
to the surface, as in Iceland, Hawaii, and New Zealand. Windmills
have undergone a renaissance: the towers of wind turbines poke up
higher into the wind and their propellers have been redesigned to
take maximum benefit of aerodynamic lift like airplane wings. But
they produce energy plentifully only where the winds blow strong
and constant.

At the end of the twentieth century, the industrialized nations
obtained only 8 percent of their primary energy from wood, hydro-,
and geothermal sources. That total doesn't include all sources—
waves and tides, for instance—but even with all these added, the
final sum at present falls far short of requirements.

"Hydrogen, hydrogen," the environmentalists chant to ward off
panic and despair. The more thoughtful corporate executives and
even a president of the United States have lately joined in. We are
told that hydrogen, the solar fusion of which has made higher life
forms possible, will enable all humans to attain and maintain a high
standard of living. That happy state will last forever because hydrogen is by far the most plentiful element in the universe. It exists in
inexhaustible quantities in the waters of our oceans.

Hydrogen burns (oxidizes) easily, producing heat but no pollutants because hydrogen combines with oxygen to make pure H_2O.
Mayor Richard Daley of Chicago once even drank a glassful of the
exhaust from one of the hydrogen-powered buses he was publi-
cizing and suffered no harmful effects. Hydrogen, like electricity,
is a wonderfully convenient means to transmit energy and, unlike
electricity, does not disappear when inactive. Store it as a gas in a
tank to inflate picnic balloons or as a liquid in the hold of a giant
freighter to empower a nation, and you can transport it where you
will and use it as you will.

There are complications, of course. Only if the ingredients are pure will the afterproduct be the same. Since purity is costly, in practice less than pure ingredients are used in hydrogen burning, leaving some pollutants—nitrogen oxide, for instance, one of the ingredients of acid rain. However, one mustn't be so insistent on the best that one rejects the better. Burning hydrogen endangers the health of humanity and of the biosphere a great deal less than burning biomass or fossil fuels.

One doesn't have to burn hydrogen to tap its energy potential, however. One can use it in pollution-free fuel cells. In these the hydrogen atoms are stripped of their electrons, which are harvested to make electric lights shine, to run computers, sewing machines, automobiles—to do work. This energy production goes on noiselessly because a fuel cell has no moving parts. The hydrogen nuclei, divested of their electrons and longing for negative companionship, meld with oxygen atoms to, again, make water. Fuel cells are too expensive now, but that may change.

There is a big worm in the manna, however. Hydrogen is a gregarious element that freely joins with other elements to form molecules and is very rarely available in pure form on earth. Considerable energy has to be spent to break it loose. Currently, the cheapest and most accessible source of hydrogen by far is the fossil fuels, which are, after all, *hydro*carbons. But using this source to cure our pollution problem is more than a little like curing a nicotine addiction by switching from cigars to cigarettes. There are other sources of hydrogen: H_2O, for instance. We can use a moderate electrical charge to split liquid water into its constituent gases, hydrogen and oxygen. The technique is accomplished innumerable times every semester for the edification of high school students in physics and chemistry classes around the world.

Electrolysis, as it is called, is a cheap and simple way to obtain a little hydrogen. Obtaining a lot—enough to power a city, a nation, an industrializing world—is a colossal challenge. The United States alone consumes 12 million barrels of petroleum a day for surface transportation: the energy equivalent in hydrogen for that amount of petroleum would be at least 230,000 tons of hydrogen. Such quantities are mind-boggling; but the problem, again, is not one of quantity. The oceans are full of hydrogen. The problem is obtaining enough electricity to break hydrogen in such enormous quantities out of the watery embrace of oxygen.

Let us again consider the case of the United States, the diva of an energy-extravagant civilization. In order to provide the electricity needed to pry loose enough hydrogen to meet its full requirements, the United States would need 400 gigawatts (400 billion watts) of electricity in addition to what it already generates. Another two hundred Hoover dams would do the job; but even if Americans were willing to accept the costs in dollars and environmental damage of that many huge new dams, there aren't rivers in size and number big enough for such structures in North America. Perhaps Americans could placard the continent with solar panels and erect forests of windmills from sea to sea. We might launch huge solar cells into orbit or deck the Moon's surface with them to transmit sun power by the thousands of megawatts down to receivers on Earth. (Perfect aim would be crucially important, of course.) Perhaps, but that solution might still fall short and is certainly not politically likely. The political problem is that very few citizens are willing to have solar panels, windmills, or other energy devices constructed in their neighborhoods. The semi-official title of the resistance is NIMBY (not in my backyard).

Perhaps we will solve our energy dilemma by doing everything mentioned in the last few pages, cobbling together a variety of environmentally friendly sources of energy. That would be complicated. The alternative is to utilize a new and very powerful prime mover that doesn't pollute. It already exists: the nuclear reactor waits at our elbow like a superb butler.

7

FISSION

So, no doubt, the more elderly and ape-like of our prehistoric ancestors objected to the innovation of cooked food and pointed out the grave dangers attending the use of the newly discovered agency, fire. Personally I think there is no doubt that sub-atomic energy is available all around us, and that one day man will release and control its almost infinite power. We cannot prevent him from doing so and can only hope that he will not use it exclusively in blowing up his next door neighbor.

—*Francis Aston, physicist and Nobel laureate (1936)*[1]

As we have seen, humans always have been "children of the sun," drawing our energy, with the exception of tiny amounts from geothermal and tidal sources, directly or indirectly from our star. But our industrial civilization requires more energy than we are at present deriving from that source. We have a job opening for an additional prime mover, for another engine that taps natural forces to do work for us, preferably without grossly polluting the atmosphere. We are at present interviewing a very important candidate: the nuclear power plant. A pound of uranium can produce as much heat as 2.5 million pounds of coal. We have developed and are continuing to develop means to harness nuclear fission and fusion, the splitting and fusing of atoms, processes seemingly more appropriate for a star than a planet. We are cultivating inti-

[1]Quoted in Richard Rhodes, *The Making of the Atomic Bomb* (New York: Simon & Schuster, 1988), 141.

macy with technologies so fruitful in energy that they could solve all of our energy problems. They could also fry us in our tracks.

As THE *fin de siècle* began, atoms were still thought to be as solid and predictable in behavior as pool balls. By the end of the 1890s, the discoveries of X rays that passed through opaque matter and of radium that fizzed with inexplicable energy swept away much of commonsense physics and startled the experts into reassessing matter, time, and space. This new vision of reality included the realization that there were stupendous amounts of energy holding atoms together, which humanity might tap for its own purposes. This fuel could fire a new and colossal prime mover.

In the first decades of the twentieth century, scientists, including Vladimir I. Vernadsky (the author who coined the phrase used for the title of this book), plus a few laymen, realized that breakthroughs in atomic physics circa 1900 were pointing the way to a new and possibly dangerous era of energy abundance. One of the more accurate prophets was H. G. Wells, the science-fiction writer, who in 1914 published a novel, *The World Set Free*, in which he raved that atomic energy would "mean a change in human conditions that I can only compare to the discovery of fire, that first discovery that lifted man above the brutes."[2] He also wrote about atomic bombs thrown by hand out of the cockpits of the primitive aircraft of the time, bombs which would produce persistent wandering volcanoes.

For decades, the two possibilities of permanent abundance and enormous destruction lingered far short of becoming probabilities. Geniuses trying to understand atomic behavior scrawled formula after formula on blackboards and then had to erase most of them. Actual experiments to prove or disprove their theories were, when possible at all, mostly tabletop affairs. There wasn't a lot of money

---

[2]H. G. Wells, *The World Set Free: A Story of Mankind* (New York: E. P. Dutton, 1914), 36.

for research on abstruse matters of so distant practical application.

World War II changed all that. In 1938, two pioneering scientists, both citizens of Nazi Germany, Otto Hahn and Fritz Strassmann, bombarded a tiny sample of uranium with neutrons (subatomic particles with no electric charge), which split uranium atoms into fragments and surges of energy, producing the first example of atomic fission to be widely recognized as such. If that phenomenon could be made self-sustaining—if a chain reaction could be initiated by which the neutrons of the splitting atoms would split additional atoms—then the puzzle of how to harness nuclear energy would be half solved. It might be possible to make a weapon powerful enough to eliminate whole cities with a single bomb apiece, or to provide all humanity with a decent standard of living (a matter of less immediate importance to world leaders in the year of the Munich confrontation of Hitler and the heads of France and Great Britain).

As German Panzer divisions rolled west and east across Europe and the Japanese fleets advanced across the Pacific, governments—especially the U.S. government—loosed whole treasuries into physicists' laps for the making of such a bomb. The first thing to do was to find out whether large-scale and controlled chain reaction was possible.

In 1942, the Italian scientist Enrico Fermi (not a Jew but married to one, an unforgivable pairing in Mussolini's country) led a project to build under the stands of a sports stadium at the University of Chicago the first nuclear power plant, or, as they are usually called, a nuclear reactor. He did so without informing Robert Maynard Hutchins, founder of the "Great Books" program of humanist studies and president of the university. Fermi judged that he just wouldn't understand, which is probably true. The language for the various aspects of the project—for instance, the term "meltdown"—hadn't been invented yet.

Their Chicago Pile One (so called because it was located in Chicago and it was what it was, a pile) consisted of 12,400 pounds of uranium metal, 80,590 pounds of uranium oxide, and 771,000 pounds of graphite as a *moderator*. (More about that in a moment.) Cadmium rods were inserted into the pile to absorb the neutrons being released spontaneously by the radioactive uranium, and thus to control the chain reaction.

On December 2, 1942, all of the rods were removed but one. Fermi ordered the withdrawal by inches at a time of the remaining rod. An instrument for sensing freed neutrons clicked and clicked faster and then settled into a steady snarl. Enough neutrons were shooting out of splitting uranium atoms to split more atoms, producing more neutrons to split yet more atoms. Self-sustained chain reaction had been achieved. The forty-three people present toasted their success with wine in paper cups. One of them phoned James B. Conant at Harvard and reported, "The Italian has landed in the New World." Conant recognized the reference to Fermi.

"How were the natives?"

"Very friendly."[3]

Atomic bombs were possible and so was the domestication of nuclear energy. Fermi's pile was a prime mover. It utilized natural phenomena to produce heat, which could be used to do work, like fire under a teakettle or a steam locomotive's boiler. So potentially hot was this first and primitive nuclear reactor that if Fermi had let the chain reaction run for an hour and a half, he would have achieved the first meltdown, confirming all the prophecies about the mad behaviors of mad scientists. He didn't because he wasn't mad. He was a splendid engineer and physicist: he held the pile for four and a half minutes to a maximum output of about enough energy for a small flashlight.

The Chicago Pile One was purely experimental. There were no means by which to cool the uranium or to harness its heat to do more than test theories and hunches; but it was the mother of all

[3] *The First Reactor*, U.S. Department of Energy/NE-0046 (December 1982), 19.

reactors, even the current giants that produce hundreds upon hundreds of megawatts of electricity.

WE SHOULD PAUSE here to try to comprehend what the giants are like. That is a challenge because of the abstruse science and engineering involved and because there are varieties of reactors. The description that follows is incomplete, but useful as an introduction to the subject.

Standard nuclear reactors have six main components. One, a *core* of fissile matter, that is, of heavy unstable atoms that are inclined to split easily, like uranium or plutonium, with rods of a substance which, when inserted into the core, smothers fission, or, extracted, permits fission. Two, a *moderator* of some kind (it had been graphite in Chicago Pile One) to slow the careening neutrons so they can be more easily captured by and destabilize (split) other atoms. Three, a *coolant* of liquid or gas to circulate around and through the core to prevent it from melting of its own heat. Four, some sort of *circulating liquid* or *gas* to take up the core's energy—the heat (in some kinds of reactors this is the coolant, too)—and pass it on to spin turbines connected to dynamos to make electricity. Five, *shields* of one kind or another to protect the workers from radiation. And six, a rugged *outer shell* to prevent the escape of radioactive dust and gases and to contain whatever explosions may occur.

There are several kinds of reactors. For our purposes, we can divide them into two categories. One, the usual kind, which takes in fissile materials and uses them until their exploitable energy is harvested. Two, the breeder reactor, which takes in fissile materials to tap them for energy, and secondarily (or in some cases primarily) to bombard non-fissile, non-weapons-grade uranium, U 238, with neutrons, and transform it into another element, plutonium, Pu 239, which is fissile and very suitable for bombs indeed.

THE HISTORY OF the utilization of nuclear energy for peaceful purposes following the holocausts at Hiroshima and Nagasaki at the end of World War II began in a spirit of rosy nuclear naïveté. Some optimists envisioned that the bombs might be used for benign purposes. Starting in 1957, the United States spent millions of dollars on the Plowshare Program for using nuclear bombs as earth movers, blowing about thirty enormous pits in the Nevada desert to prove that they could serve as colossal bulldozers. Americans toyed for years with a plan to blast a new sea-level Panama Canal (radioactivity be damned!), finally dropping the idea because of the opposition of the "prospective host country." The Nixon administration let the Plowshare Program fade away in the 1970s.

There were more practical plans to use nuclear reactors to provide the world with all the electricity it might need. Lewis L. Strauss, chairman of the U.S. Atomic Energy Commission, which supervised nuclear affairs for America for a quarter century after World War II, proclaimed that nuclear reactors would soon be producing electricity so cheap that it wouldn't be worthwhile to meter it.

In 1951, an American successor to Fermi's Pile (called a reactor by this time) produced electrical power from nuclear heat, enough to light four light bulbs. In 1955, the United States launched the *Nautilus*, a submarine powered by a small and amazingly powerful nuclear reactor. It did not burn up its fuel in a few weeks and did not need air for its combustion. The *Nautilus* could cruise for month after month without refueling and without surfacing. During 1958, it passed under Arctic ice and surfaced at 90 degrees north, the first vessel ever to do so. The captain proclaimed, "For the world, our country, and the Navy—the North Pole." By then the Soviet Union already had a nuclear-powered icebreaker, the *Lenin*, and soon it added schools of nuclear submarines to match America's. In 1961, the U.S. Navy commissioned the world's largest ship, the USS *Enterprise*, a nuclear-powered aircraft carrier touted as able to cruise at 30 knots and for up to 400,000 miles without refueling.

Not all the advances were directly associated with the cold war between the United States and the USSR. The Russians opened the first civilian nuclear power station at Obninsk not far from Moscow in 1954. The following year, the village of Arco, Idaho, population 1,000, became the first town powered with electricity from a reactor. In 1957, the world's first large-scale reactor began producing electricity at Calder, England, for public and private consumers. A year later an even bigger one began operation in Shippingport, Pennsylvania; and France, Japan, Canada, and most of the world's industrialized nations were scrambling to build their own reactors.

A sort of desperate optimism about nuclear power swept round the world. President Dwight D. Eisenhower delivered an "Atoms for Peace" speech before the United Nations. "A single air group," he warned in 1953, can now deliver "a destructive cargo exceeding in power all the bombs that fell on Britain in all of World War II." But, he continued, there was an alternative to a world devastated by nuclear power: a world saved by nuclear power.

The United States knows that peaceful power from atomic energy is no dream of the future. That capability, already proved, is here now—today. Who can doubt, if the entire body of the world's scientists and engineers had adequate amounts of fissionable material with which to test and develop their ideas, that this capability would rapidly be transformed into universal, efficient and economic usage?

The Noble laureate and new chairman of the Atomic Energy Commission, Glen Seaborg, proclaimed in 1971 that by the end of the century half his country's needs for electricity would be supplied by nuclear energy and nuclear-powered spaceships would be carrying humans to Mars. The United States already had twenty-two commercial nuclear power plants in operation. They supplied only 2.4 percent of the nation's electricity, but there would obviously be more of them soon and America had already put the first nuclear reactor into space. But in the 1970s enthusiasm waned. The public demand for electricity had not increased as rapidly as had been predicted, and actual experience in dealing with radioactivity encouraged more regulation, more safety equipment, more training for personnel, more halts for inspections—all of which cost a lot more money. And there were mishaps that cooled popular ardor.

The great majority of nuclear accidents in the world were minor, and remedied effectively by such procedures as simply reinserting the control rods. There were means to do this automatically: for instance, when electricity failed, the rods, suspended over the core electromagnetically, were released to fall into the core and smother chain reaction. In America the procedure was called the Safety Control Rod Adjustment Mechanism, usually referred to by the discomforting acronym SCRAM. At that time SCRAMs occurred several times a year in most reactor plants.

The one characteristic about fissile elements of which everyone was conscious, even paranoid, was that they could, as bombs, detonate with almost unimaginable force, killing people by the scores of thousands. That anxiety was extended to nuclear reactors, but not justly. Their threat was never an instantaneous chain reaction and a nuclear explosion, because reactor fuel is not as richly fissile as the contents of a nuclear bomb.

But nuclear explosions are not the only kind of explosions. Reactor fuel, fissile or not, is intensely radioactive; it is capable of producing immense heat to ignite or melt or disrupt in one way or another quite nearly anything, including itself, with appalling consequences. It can turn water into steam that expands so rapidly that the effect is of a gigantic exploding boiler. Under some conditions it can divide water into oxygen and hydrogen, elements for chemical explosions.

Melodramatic scenarios for reactor accidents always climax with the so-called China Syndrome. This is a meltdown in which the molten core melts right down through the floor of the reactor building and on and on—straight through to China, so to speak. When the sizzling core hits the groundwater, it instantly turns it to steam, setting off a terrific explosion, spraying radioactive materials far and wide, killing numbers of people. A runaway meltdown would create what today we would call a "dirty bomb."

In the 1950s and 1960s there were a number of accidents in nuclear power plants worldwide. At Windscale in England, the graphite around the uranium fuel caught fire in 1957 and there was a partial meltdown. The offending reactor was flooded with water and the fire extinguished, but the farms nearby were contaminated with radioactivity. A few months later there was an accident involving nuclear wastes at Chelyabinsk in the Soviet Union. According to the rumors that eventually trickled through to the West, many had died; but that was all uncertain and far off in a strange land. In 1969, a reactor at Saint-Laurent, in France, one of the most powerful in the world, had a partial meltdown. There were other accidents as well, but governments didn't publicize them and the public as yet wasn't much interested.

The U.S. Atomic Energy Commission was concerned enough to spend $4 million for an assessment of nuclear reactor safety. The multivolume report, known as WASH-1400 or the Rasmussen Report—named for the MIT professor who commanded the staff of fifty that produced it—came out in 1974. Its estimate was that if everything that possibly could go wrong at a nuclear reactor site did so, leading to a meltdown of the core, then about 3,300 people would die; there would be $14 billion in property damage; and inhabitants in a 290-square-mile area would have to be evacuated. That would be bad, very bad, but not the end of life on the planet or even of civilization. And it wouldn't happen, anyway. Professor Norman Rasmussen announced that the chances of being killed by a nuclear accident were no higher than those of being hit by a meteor.

On March 28, 1979, an accident at one of the reactors at Three Mile Island near Middletown, in Pennsylvania, closed the era of rosy optimism about nuclear power. The accident began when the coolant flow stopped, various mechanisms malfunctioned, and the operators lost track of what was happening. There was enough uncontrolled radioactivity to make heat for explosions and a meltdown. The interaction of melting metals (including 50 percent of the core) and the coolant steam produced something unforeseen: a large hydrogen bubble. It could have exploded, conceivably splitting the containment dome and spreading radioactive dust over considerable areas of the thickly populated East Coast of the United States; but it didn't. Some radioactive materials escaped into the atmosphere, but most stayed put. There were no immediate casualties or, according to careful investigations years later, any in the long run. Thousands of people in the vicinity of the reactor were evacuated. The cleanup took years and cost millions of dollars.

Opposition to nuclear power plants soon became nearly as widespread as opposition to nuclear bombs. Thousands of anti-nuke protestors marched through Washington, D.C. A lurid movie, *The China Syndrome* (1979), starring Jane Fonda and Jack Lemmon, spurred more opposition. The U.S. Nuclear Regulatory Commission held hearings on emergency core cooling systems that lasted over twenty-three months and filled 22,000 pages.

If what happened in 1986 at Chernobyl, Ukraine—then part of the Soviet Union—had been in a movie, we might remember it as a campy comedy of errors. But what did happen was real and much worse than the accident in Pennsylvania in its effects, cost, and influence on public opinion everywhere. Reactor No. Four at Chernobyl was poorly designed. It did not have a containment structure and was operated by badly trained technicians, supervised by officials accustomed in a totalitarian society to suppress bad news.

On April 25, 1986, the technicians shut down several emergency systems to test the stability of this reactor in trying circumstances. They proceeded not knowing that the backup cooling system, which should have been operating continually to prevent meltdown, was not operating. There was an enormous surge of heat, which interacted with the water that was present to create huge explosions, tossing off the 1,000-ton roof of the reactor, blowing radioactive materials out to the surrounding region and high into the atmosphere.

Firemen rushed in to control the conflagrations at the ground level while helicopters flew over what had been a reactor and now was a volcano, dropping 5,000 tons of various materials to quench its flames. The fires lasted for days. Firefighters and airmen died of exposure to radioactivity. There were thirty immediate human casualties. The Soviet government evacuated 130,000 people from Chernobyl and its vicinity and resettled them elsewhere.

The immediate damage done to those in the vicinity of the meltdown is easily measured. The long-range effects of the radioactive fallout are complicated, often delayed, and can be subtle. Nor is it clear that Soviet and Ukrainian statistics, particularly those pertaining to failures, are dependable. The U.N. Scientific Committee on the Effects of Atomic Radiation estimated the death toll from the Chernobyl accident might turn out to be about 1,000; the Ukraine Radiological Institute suggested 2,500. Estimates of the number of victims who will eventually die of cancers caused by the fallout range as high as 475,000. The thyroid cancer rate for children in Chernobyl and its vicinity fifteen years old and under jumped from 4 to 6 per million in 1981–85 to 45 per million in 1986–97. V. K. Savchenko, a geneticist from Belarus, which received more fallout from the Chernobyl accident than any other nation, rates it as "the greatest technological catastrophe in human history."[4]

That may well be, but we don't know whether or not it was the greatest technological killer of all time and probably never will. Pregnant women who were at or near Chernobyl in 1986 may have given birth to less healthy offspring than otherwise would have been the case. We won't know until those children live out their lives—if then. Radioactivity can suppress the immune system; then the victim dies not of radiation but of pneumonia or measles or some other infection.

---

[4] V. K. Savchenko, *The Ecology of the Chernobyl Catastrophe* (Paris: UNESCO, 1995), xv.

In the eighties, after Three Mile Island and then Chernobyl made their debuts on prime-time news, the nuclear energy industry, which had already been losing momentum, lost many supporters and acquired many more detractors. In 1967, it took sixteen months to build a standard nuclear power plant in the United States; in 1972, thirty-two months; in 1980, fifty-four months. An American nuclear power plant that in the early seventies cost $170 million, cost $1.7 billion a decade later, and by the late eighties, after Chernobyl, $5 billion. At the end of the 1970s there were seventy commercial nuclear power plants in the United States, but not one more was ordered after 1978. Most of those already far along in planning and construction were finished, and the United States, arguably the homeland of nuclear energy for peaceful purposes, ended the century with just over one hundred plants, a little less than a quarter of the total for the world. And that, at least for the time being, was that.

The saga of the projected plant at Shoreham on Long Island, New York, neatly illustrates the ascent and descent of the reputation of nuclear power in American public opinion. Construction of an 810-megawatt plant began in the optimistic year of 1970. Environmentalists objected from the beginning, and after the Three Mile Island imbroglio crowds joined them to demonstrate against the new plant, but construction continued.

After the Three Mile accident the federal authorities required the framing of plans for the evacuation of everyone within ten miles of a nuclear power plant in case of trouble. That area, twenty miles across, was officially recognized as a danger zone. Long Island, 118 miles long by 12–20 miles wide, had a population of about 5 million, with Shoreham located approximately halfway from one end of the island to the other. The people to Shoreham's east had no bridge in their vicinity to the mainland and could not flee a nuclear accident at the new power plant on the existing road system without passing through the danger zone. They would be marooned at their part of the island, waiting to fry, said the worriers. After years of wrangling, the Shoreham plant was canceled in 1989. It had cost $5.5 billion.

THE UNITED STATES is fortunate in that it has, at least for a generation or two, a relatively easy choice of energy sources. It has lots of fossil fuels within its borders or nearby and doesn't necessarily have to play dangerous games with uranium and plutonium to keep the lights on. Germany, Sweden, and Great Britain, other nations with their own coal, have also decided to wait and see.

France, which has little fossil fuel, had fewer options and also traditions of grandeur. After the war, Charles de Gaulle, anxious that his nation should have its own nuclear bomb, the *force de frappe*, also wanted his people to take the lead in peaceful development of nuclear power. France, whose acumen in atomic science stretched back to Marie and Pierre Curie, pioneers in the study of magnetism and radioactivity, was well prepared to take the lead. Japan, Taiwan, and South Korea, also lacking in fissile materials, followed the French, not the American, example.

In the year 2000 there were well over four hundred reactors in more than thirty nations throughout the world. France obtained nearly 80 percent of its electricity from its nuclear power plants and even exported some. South Korea got 40 percent of its electricity and Japan almost 35 percent from nuclear reactors. The United States and Great Britain, the first two nuclear powers, obtained only a fifth of theirs from that source.

THE COAL-RICH nations have not committed themselves to nuclear power plants for their electricity because theirs is not a matter of "No oil, no coal, no choice." But for how long can they hold back? The usual answer is for a long time—longer probably than any of us will live. But that, dear reader, is well short of forever. The demand for electricity in the developed societies increases. In 1990, nuclear power plants were supplying Americans with only 20 percent of their electricity, but that was more kilowatts than all sources had supplied in 1956. The demand of the peoples of the developing societies—China, Brazil, India, and others—soars as they hope to achieve industrially in a generation what the West achieved in a century. Sooner or later, all nations, initially rich in fossil fuels or not, will face the problem of energy shortages.

There seem to be three paths to pick from. One, we can return to an energy-impoverished existence such as we all lived in circa 1750. Two, we can choose to practice energy sanity. We can limit population growth and material acquisition, live frugally, and draw our energy from the sources—solar, hydro, geothermal, wind—considered in chapter 9. The first choice is, put simply, not available to the 6 billion-plus of us presently alive because there is no way that an eighteenth-century economy can support that many people. The second smacks of celibacy and dieting, neither of them human fortes. In the third, we can follow France into the possibly radioactive mists. Let us proceed with brief summations first of the positive arguments for nuclear energy and then of the negative.

To begin with, radioactivity is not a demon crouched in the bushes ready to leap out at us, but simply a fact of nature. Radioactivity in small doses is everywhere. Even our bodies are radioactive. It streams in from space and seeps out of the ground. The chief source of the radioactivity to which we are exposed is not the nuclear reactor, but our world and universe. Of course, radioactivity in excess should be avoided, but it is not anathema. It is natural.

Electricity from nuclear reactors is cheaper than that from fossil fuel plants, if you include in your calculations something economists usually ignore: the costs of the fossil fuel energy in smoke and smog on human lungs, plants, and global temperatures. Nuclear energy is, day after day, year after year, clean. Yes, the radioactivity in nuclear power plants is dangerous, but it has *very* rarely escaped in more than minuscule amounts. We expose ourselves to more radioactivity by burning coal, which contains minute quantities per ton.

The accidents at Three Mile Island and Chernobyl are frightening, but, examined objectively, they have aspects that can be seen as encouraging. Many things went wrong at Three Mile Island, but no one was killed at the time and twenty-five years later mortality related to the accident still appears to be zero. The average dose of radioactivity received by people within five miles of the accident was no greater than that of a chest X ray. Nearly everything went wrong at Chernobyl and the fallout, which was great enough to kill a stand of pines ten or so kilometers away, spread over most of Europe. Thirty-one people died immediately or very soon after, and thousands more deaths were predicted. But a conference sponsored by the European Union, the World Health Organization, and the International Atomic Energy Commission ten years after the disaster concluded that "apart from thyroid cancer, there has been no statistically significant deviation in the incidence of other cancers attributable to radiation exposure due to the accident. In particular, to date no consistent attributable increase has been detected in the rate of leukemia, one of the major concerns of radiation exposure."[5] And only three of the several hundred children with thyroid cancer had died thus far.

As for the widely credited threat that terrorists will seize nuclear power plants and somehow wield them as mega-weapons, we now know how to build nuclear reactors that are "meltdown proof," that will cool themselves automatically if madmen try to use them to kill or even if their attendants innocently fall asleep on the job.

The number of people who have died of radiation since the Hiroshima and Nagasaki bombings is tiny compared to the number of those killed by coal smoke and automobile exhaust. These old-fashioned emissions—one could almost say traditional emissions—escape our censure because they have been a part of our lives for generations and because they kill continually at a rate of only a few

[5]Nuclear Energy Institute, "The Chernobyl Accident and Its Consequences" (July 2000). http://www.nei.org/doc.asp?docid=456. (Viewed Aug. 31, 2004.)

people in a day. In America in the twentieth century, accidents in coal mines killed about 100,000 miners and injured 4 million. In 1969, 50,000 were suffering from black lung disease. We know that the number of other people who died of lung cancer, emphysema, and other diseases attributable to coal smoke is much greater, though difficult to disentangle from cigarette smoke.

In 2002, U.S. Secretary of Energy Spencer Abraham, echoing the nuclear power optimists of the 1950s, proclaimed that he was looking forward to the day when nuclear reactors might supply all of us with energy that is "safe, abundant, reliable, inexpensive, and proliferation resistant."[6]

THERE IS AN opposing view, of course. Electricity gained by harnessing nuclear power is *not* cheap if you include the full costs in its price, for instance, the expense of shutting down nuclear reactors. All must be retired after forty years at most, their structures, radioactive to some degree, disassembled, their soils scraped up, and all put safely aside somewhere. Years go by before the sites can be opened for new use: the decommissioning of the Fort Vrain Nuclear Generating Station in Colorado began in 1990, took six years, and cost $188 million.

We must not build nuclear reactors because their fuel and, in the case of the breeder reactors, their products are much too tempting to terrorists. They might steal the fissile materials and make bombs—fission bombs to kill hundreds of thousands if they are skillful enough or dirty ones to kill tens of thousands if they aren't. The danger of such attacks is greater today than at the peak of the cold war.

As for the number of deaths attributable to the Three Mile Island and Chernobyl accidents, we dare not take comfort from their alleged low totals until the generations of their victims have

[6]Declan Butler, "Nuclear Power's New Dawn," *Nature*, 429 (May 20, 2004), 238.

lived out their lives. You can die at sixty years of age of radiation exposure suffered at five; furthermore, the dire effects of that exposure may not appear for another generation. And Three Mile Island and Chernobyl were not the final accidents. For instance, there was one in Tokai-mura, Japan, in 1999 that released enough radiation to kill two workers and injure thirty-odd more. Except for Chernobyl, we have been lucky so far, but a truly calamitous accident is inevitable if we continue to live by and with reactors.

The problem of what to do with the radioactive discards from our nuclear power plants—the uranium and plutonium and everything sufficiently exposed to their radiation to become radioactive—is insoluble. Some of these materials will be briefly dangerous; others will be deadly for more years than have passed since the last Neanderthal walked the earth. The claims that we may be able to process the most dangerous radioactive wastes so they will be deadly only for centuries instead of millennia do not delight. We can bury radioactive detritus thousands of feet down in dry rock chambers, but we cannot guarantee that they will never threaten the lives of human beings again.

In July 2004, the U.S. Court of Appeals for the District of Columbia ruled that measures recommended by the federal government for protecting the public from radiation leaks at the Yucca Mountain atomic waste depository for ten thousand years ("two Abrahams ago," as one expert put it, referring to the Old Testament patriarch) were inadequate because that was too brief a time. The National Academy of Sciences helpfully suggested that 300,000 years might be more appropriate.

THE ADVOCATES of nuclear power propose that humanity will be careful enough to avoid the chasms that skirt the path of energy abundance. The opponents of nuclear power seem to be calling for frugality, self-control, and for a great many minor, but in total accumulation enormously difficult and burdensome, social and technological changes.

Making the choice requires sober consideration of matters that arouse extreme anxiety and that very little in our evolution or, certainly, our experience since the Paleolithic has prepared us to understand. We are obliged to reason dispassionately about radioactivity, which our five senses cannot detect until we smell our hair beginning to fry, and about substances which, in bombs, have destructive powers comparable to those of volcanic eruptions. We are required to think of these spatially in terms of the entire biosphere and temporally in terms of periods that are longer than our species' existence. And we are also required to assess what our species will do to the biosphere and to itself if its energy greed is not satisfied by nuclear power.

This subject, which arouses anger, fear, and panic, requires cool and careful analysis.

Coda: Chernobyl in the USSR and Reindeer Meat in Sweden

In the spring of 1986 the Chernobyl accident put the 4 million people of Belarus, Ukraine, and contiguous Russia in immediate jeopardy, but in general the danger diminished as time passed and the fallout rarefied as it drifted farther away from its point of origin. In America, its radioactivity could be measured only by delicate instrumentation. The dispersal was not, however, perfectly even. For instance, radioactive traces of Chernobyl are particularly notable in the Alps and Greenland and will be useful in dating levels of glacial ice.

Unfortunately, the benign effects of time and distance could be reversed, as may have been the case in northern Scandinavia and Russia, where more than sixty thousand Sami (often called Lapps) lived. They might be called the reindeer people because for a millennium and more, reindeer have been as much the center of the local way of life as cattle are for the Masai of East Africa. Change is on the way, of course: young Sami go to universities and their parents herd their livestock on snowmobiles; but the truth endures that their diet, clothing, tools, and living patterns are still in part or whole associated with reindeer. In Sweden in 1986 there were about 20,000 Sami and 270,000 reindeer.

Reindeer normally don't live in barns and feed on what their owners provide, but roam through the taiga browsing on plants, berries, and mushrooms. They are, especially in winter, largely dependent for nourishment on reindeer moss. It has no roots or stems and derives moisture and nourishment from the air, which in 1986 was rife with Chernobyl fallout.

The moss quickly absorbed the radioactive substances and concentrated them. The reindeer ate the lichen and concentrated the radioactivity again in their meat and bones. The animals did not fall sick, but scientists and governments and the popular press decided their meat was dangerous. Sweden's Food Administration issued a warning that reindeer meat should not be eaten. The market for reindeer meat at home and abroad collapsed.

Eighty percent of Sweden's reindeer were slaughtered in the year following the first detection of Chernobyl fallout in Sweden. Some were buried in pits under signs reading: DANGER: RADIOACTIVE MATERIAL UNDERGROUND. Some of the meat went to mink farms, the minks being destined for the furrier and not the grocery store. The Swedish government has paid the equivalent of tens of millions of dollars for testing and slaughtering the deer in the years after Chernobyl.

Paradoxically, there is little evidence among the Sami of lasting physical injury associated with the Chernobyl fallout. In the 1990s, radioactive contamination of reindeer moss decreased at a rate of 15 percent a year. In 1986–87, when 95,554 of Sweden's reindeer were slaughtered, only 20 percent of their meat was considered fit for human consumption. By 1997, 99 percent was considered fit.

Birgitta Ahman of the Reindeer Husbandry Unit at the Swedish University of Agricultural Sciences in Uppsala assures us that the level of radioactivity in reindeer meat declined sharply after 1986–87. A dozen years later it was down to one fifth or so of the first post-Chernobyl winter levels and in some places all the way down to pre-Chernobyl levels.

There seems to be no reason to avoid reindeer meat; the market for it has revived, and the reindeer herders and their herds still roam the taiga and tundra.

8

FUSION

. . . That snare that shall catch the sun.

—H. G. Wells, novelist (1914)[1]

The Promethean quest to harness fusion energy, to capture the fire of the Sun on Earth, has sometimes been called the greatest technological challenge of all time.

—T. Kenneth Fowler, physicist (1997)[2]

O f atomic energy zealots, none have soared higher in hope and confidence than the advocates of hydrogen fusion— the fusing, not splitting, of atomic nuclei to transform matter into energy. In 1955, at the first International Conference on the Peaceful Uses of Atomic Energy, Homi Bhabha, India's most prominent nuclear physicist, predicted that "a method will be found for liberating fusion-energy in a controlled manner within the next two decades. When that happens, the energy problems of the world will truly have been solved. . . ."[3] A half century later, hydrogen fusion is still not domesticated and our energy problems loom larger than ever. So many years have passed since Bhabha's optimistic outburst that, as one government report puts it, "The

---

[1]H. G. Wells, *The World Set Free: A Story of Mankind* (New York: E. P. Dutton, 1914), 15.

[2]T. Kenneth Fowler, *The Fusion Quest* (Baltimore: Johns Hopkins University Press, 1997), ix.

[3]Quoted in Helge Kragh, *Quantum Generations: A History of Physics in the Twentieth Century* (Princeton: Princeton University Press, 1999), 292.

fusion community is aging. . . ."[4] Young scientists, it would seem, are reluctant to bet their careers on fusion.

THE SUBJECT OF hydrogen fusion brings us full circle to where we began, at the center of the Sun, where the headlong collision of hydrogen nuclei (protons) has been happening without our help for billions of years and will continue to do so for billions more. As sketched in chapter 1, the energy that is a side effect of the resulting nuclear fusion flows out of the Sun and makes our lives possible here on Earth. The solution to our energy problems is simple: all we have to do is domesticate hydrogen fusion just as we did fire a long while ago.

The rewards would be tremendous. We need more energy, a lot more energy, and a gram of hydrogen converted to helium releases about 100 million times as much energy as a gram of TNT and eight times as much as a gram of uranium. We need an energy source that exists in superabundance. Fossil fuels may possibly last for generations more, but will be increasingly difficult and expensive to secure. The natural deposits of uranium will also run out someday. We can produce eminently fissile plutonium with breeder reactors, but it may well be the most persistently dangerous element of all, with a half-life (the amount of time in which the radioactivity of a given substance decreases by 50 percent) of over twenty thousand years. Our oceans are an inexhaustible source of hydrogen and even of the relatively rare deuterium, the kind of hydrogen with a nucleus of one proton plus one neutron. This hydrogen isotope works better for fusion than the more common one-proton kind. There is the energy equivalent of 300 gallons of gasoline in the deuterium in every gallon of seawater.

We need an energy source that won't endanger us the way nuclear

---

[4]http://books.nap.edu/books/N1000487/html/1.html. National Academies, National Research Council, Letter Report, Burning Plasma Assessment (Dec. 20, 2002), 2.

fission does. Meltdown is impossible with fusion. When fusion halts, heat production begins to fall. Fusion, compared to fission, produces only some radiation as a by-product. Some tritium—the radioactive isotope of hydrogen with a nucleus of one proton and two neutrons—is created, but in small quantities. Moreover, tritium does not accumulate in the human body and has a half-life of only a dozen years. Retirement of fusion reactors would be easy: the reactor equipment and nearby structures would be radioactive, but for only a few years. There would be no necessity for underground tombs marked with sculptured skull and crossed bones to scare off our distant descendants from hydrogen's radioactive by-products.

We need energy plants that will not contaminate our air, waterways, oceans, crop fields, and playing fields with carbon dioxide and other pollutants. Hydrogen fusion plants wouldn't. Helium, the chief by-product of hydrogen fusion, is harmless and can be released safely into the atmosphere.

So, why haven't Bhabha's colleagues—or the Chinese or the French or Russians or Japanese or Americans or somebody—domesticated hydrogen fusion yet?

To fuse hydrogen nuclei, which are positively charged, we have to overcome their enormous mutual repulsion. That is possible in the Sun's center because of its extremes in temperature and pressure—temperatures of millions of degrees and gravitational pressures powerful enough to contain a perpetual hydrogen bomb explosion. The trick is to produce comparable conditions here on Earth, with the key requirement a temperature of many millions of degrees. At such a temperature and with proper confinement, hydrogen fusion can become self-sustaining. This is called "plasma-burning," meaning sustained hydrogen fusion, the fusion equivalent of a chain reaction.

We manage this for a nanosecond or so when we use a fission bomb to trigger a hydrogen bomb. The fissioning provides the necessary heat, and so swiftly does it do so that inertia holds the hydrogen atoms in place long enough—the merest sliver of time—for fusion to happen. The Soviet Union once produced a hydrogen fusion explosion 4,600 times more powerful than the Hiroshima bomb. Impressive—but still not a practical means for boiling water for tea. For everyday purposes we have to learn how to scale down fusion so that it happens to tiny amounts of hydrogen at a pedestrian pace. Our situation is like that our distant ancestors would have faced if they were trying to learn how to make campfires not with wood, but with sticks of dynamite.

How do you domesticate the H-bomb? An attempt to answer that question requires consideration of some of the most recondite efforts on the far frontiers of contemporary physics. I will proceed from the least to the most promising effort. The efforts that have attracted the most attention are cold fusion, sonoluminescence (sound-light), inertial confinement, and magnetic confinement.

THE SCIENTIFIC COMMUNITY and the wide world in general first learned about cold fusion at a press conference at the University of Utah on March 23, 1989, attended by two hundred reporters and television crew members from the BBC, Reuters, and others. The miracle makers were two reputable electrochemists, Professor B. Stanley Pons of the University of Utah and Professor Martin Fleischmann of the University of Southampton. They announced that they had achieved sustained fusion of hydrogen nuclei *at room temperature*. Their experiment, they said, produced four times as much energy as they had introduced to initiate the process; indeed, so much energy that an explosion had occurred, destroying a cupboard and gouging a divot in a concrete floor.

That night Dan Rather, anchorman of CBS's popular evening television news, opined that the two professors might have accomplished "a remarkable breakthrough." The next day the *The Wall Street Journal* featured them on page one under a headline that read "Taming the H-bombs?" *BusinessWeek*, *Newsweek*, and *Time*, three of the most widely read magazines in the United States, soon after published cover stories on what was now universally known as "cold fusion." The miracle received similar coverage in the popular press around the world. President Chase Petersen of the University of Utah predicted that cold fusion might "rank up there with such things as the invention of fire."[5]

The president of the United States took notice and called top scientists to Washington to confer. The U.S. Department of Energy assembled a panel of renowned fusion experts to advise the government. The House of Representatives' Committee on Science, Space, and Technology held a hearing about "test tube fusion" in which Pons, Fleischmann, University president Petersen, and business consultants participated. For starters, Petersen wanted $25 million in federal funds for cold fusion research and several times that sum to found an institute to proceed on the basis of the Pons-Fleischmann experiment.

That experiment was simple, inexpensive, and elegant. It was a version of electrolysis. Pons and Fleischmann's innovations were two in number. First, they didn't use plain water but "heavy water," with its weighty hydrogen isotope, deuterium. Second, they used palladium for the terminal attracting hydrogen, a metal gluttonous for hydrogen. During electrolysis, it packs hydrogen atoms into itself like a sponge absorbing water, only much more densely. The experimenters were sure that the hydrogen atoms crammed together so densely within the palladium that the mutual repulsion of their nuclei wouldn't keep them apart and fusion would occur, releasing energy.

As reports of the March press conference spread across the world, scientists everywhere snapped to attention. Edward Teller, "father of the H-bomb" and one of the world's most brilliant nuclear physicists, was enthusiastic. Physicists and chemists in North America, Europe, Japan, India, and elsewhere tried to replicate the Pons-Fleischmann experiment. Many were able to—or thought they were

[5]Eugene F. Mallore, *Fire from Ice: Searching for the Truth Behind the Cold Fusion Furor* (New York: John Wiley & Sons, 1991), 56.

able to. But scientists at prestigious institutions like the Harwell Laboratory in England, Massachusetts Institute of Technology, and the California Institute of Technology, did not. Of course, they couldn't be sure they were exactly replicating the Pons-Fleischmann experiment because thus far the chemists had not published a paper precisely laying out the details. Concern about patent claims and the potential profits to be made may have overruled normal scientific behavior. As President Petersen put it: "Upon the advice of our patent counsel, it is not possible for the University of Utah to share research results with other laboratories, particularly national laboratories, until the information has been incorporated into patent application and the application is on file in the patent office."[6]

Under such circumstances many attempts at replication were bound to stumble, but so very many of them produced results so unlike those of the Pons-Fleischmann experiment that doubts accumulated. Further, if the first cold fusion experiment had produced so much energy that an explosion occurred, then why wasn't there anything remaining of the radiation that would accompany such an explosion? After CNN television showed a picture of a reporter so close to a glass bulb containing the active elements of a cold fusion experiment that he seemed to be touching it, one scientist sarcastically expressed surprise that none of the man's body parts had fallen off.

In May 1990, *Nature*, one of the most prestigious of all the scientific journals, published several highly critical articles about cold fusion. Enthusiastic rooters for cold fusion at Texas A&M University and other institutions retracted their reports of having replicated the Pons-Fleischmann experiment. The Department of Energy panel dealt the *coup de grâce*: "Nuclear fusion at room temperature, of the type discussed in this report, would be contrary to all understanding of nuclear reactions in the last half-century."[7]

[6]John R. Huizenga, *Cold Fusion: The Scientific Fiasco of the Century* (Rochester, NY: University of Rochester Press, 1992), 55.
[7]Ibid., 245–46.

The scientist and journalist Frank Close commemorates the cold fusion episode in its full duration as "the most bizarre 500 days in the history of modern science."[8] The president of the University of Utah retired under a cloud. Pons and Fleischmann, at last notice, were living and working unobtrusively in Europe.

SONOLUMINESCENCE AS A candidate in the hydrogen fusion race first showed up in the mid-1990s, when it attracted a good deal of scorn, largely because it reminded people of recently discredited cold fusion. In March 2002 it resurfaced, this time as the subject of a paper in the prestigious journal *Science*.

The basic procedure is easy to describe in general terms: you achieve sonoluminescence by means of cavitation, a phenomenon that attracted attention when it was discovered that poorly designed watercraft propellers can agitate water so vigorously as to wrench its molecules away from each other, producing pockets of near vacuum. These bubbles collapse instantly and violently, producing flashes of light and, apparently, very high pressures and temperatures. Cavitation can be powerful enough to make pits in the metal of the propellers.

Researchers in sonoluminescence with hydrogen fusion in mind are using very high frequency sound waves to produce cavitation in heavy water. Cavitation in heavy water may produce a spray of neutrons and the creation of trace amounts of tritium (the radioactive isotope of hydrogen), the result of something akin to fusion. The physicist Rusi P. Taleyarkhan, of Purdue University and Oak Ridge National Laboratory, has maintained that "For the first time in history, we've been able to use simple mechanics to initiate and control nuclear forces."[9]

Sonoluminescence might conceivably be exploited to trigger

---

[8]Frank Close, *Too Hot to Handle: The Race for Cold Fusion* (Princeton: Princeton University Press, 1991), 326.

[9]P. Weiss, "Bubble Fusion," *Science News*, 165 (March 6, 2004), 149.

fusion in hydrogen on a large scale, but thus far it is a matter of tiny, tabletop experimentation with an apparatus no bigger than a stack of three coffee cups. We also do not yet understand what exactly is happening.

INERTIAL CONFINEMENT IS superficially easy to visualize. It is a way to produce very small thermonuclear detonations in confined quarters on schedules of our choosing. That seemed and probably was impossible until the invention in 1960 of the laser, a device that emits a very narrow beam of light, which, if desired, can be intense enough to etch metal. In successful inertial confinement experiments, BB-sized glass beads containing deuterium and tritium have been fried by pulses of light, X rays, or atomic particles from dozens of converging beams. The jolt, billions of watts per pulse (sometimes for a nanosecond in excess of the total output of all the power plants in the world), implodes the beads. The resulting pressure and temperature are comparable to those at the center of our Sun. Hydrogen fusion happens instantly. The beads are briefly very tiny stars.

Jeff Quintenz, director of the Pulsed Power Sciences Center of the Sandia National Laboratories in New Mexico, compares inertial confinement fusion to the functions of an internal combustion engine: "Squirt in a little bit of fuel, explode it. Squirt in a little bit of fuel, explode it."[10] For this technique of fusion to become a practical way of producing energy in quantity, the hydrogen beads will, so to speak, have to show up at *exactly* the center of the converging beams' bull's-eye at *precisely* the right time several times a second, and the debris from every detonation will have to be carried off before the arrival of the next bead. And somehow a sort of self-sustaining chain reaction will have to be achieved. These challenges are daunting, but many of Nikolaus

[10]Kenneth Chang, "New Fusion Method Offers Hope of New Energy Source," *New York Times*, April 8, 2003.

Otto's contemporaries must at first have thought the same about his engine.

MAGNETIC CONTAINMENT HAS won more approval and government appropriations than the other techniques that might lead us all to the promised land of hydrogen fusion power. It assumes that fusion must occur here on Earth as it does in a star, that is, suspended in space. The fuel is deuterium and, often, tritium, so hot that their atoms have broken apart into separate subatomic particles, forming a sort of vapor that will transmit electricity. This plasma suspends in a vacuum chamber like a star in space. If all goes well, it touches nothing and thereby maintains its temperature and is available for manipulation.

Run an electric current through two parallel wires in the same direction and they will tend to draw together magnetically. Run a current through plasma and you get the same "pinch" effect: the plasma draws together into a sort of free-floating cylinder. Unfortunately, plasma in this shape leaks heat at both ends and, even worse, is not stable. It kinks, touches the walls of the chamber, cools, and the experiment fails before we can try to initiate fusion.

So, bend the cylinder around into a doughnut. Inside the body of the doughnut (not inside the hole in the middle), the plasma particles circle in corkscrew spirals (following magnetic lines like iron filings spread on a sheet of paper over a magnet) that feed into themselves, round and round, confining the plasma. This limits leakage, making it easier to drive the plasma to higher and higher temperatures, high enough, ideally, to initiate fusion, as in the core of the Sun. The plasma is held in place by enormously powerful magnets, to which plasma reacts not by being attracted but repelled and compressed. The name of this structure is *tokamak*, an acronym of the Russian words for a toroidal (doughnut-shaped) magnetic chamber. The Russians invented it in 1968 and the world has been following their lead every since.

Inside the tokamak, the hydrogen plasma can be subjected to

increased electrical currents, magnetic compression, radio frequencies, and injections of atoms to create temperatures of many millions of degrees, driving hydrogen nuclei to such frantic motion that they cannot avoid banging heads, and then fusion occurs. But perfect isolation of the plasma is difficult to maintain, the temperature sinks, and again and again researchers have found themselves in the situation of a man burning wet wood with a blowtorch.

Enormous advances in understanding plasma behavior have been made since Homi Bhabha's day. New and better computers enable us to model plasma behavior more and more accurately. We are now able to measure in detail the characteristics of plasmas at temperatures ten times that of the surface of the Sun. We are getting closer and closer to the magic breakeven moment when experiments will produce as much energy as invested—when output will equal input, when the plasma will "burn." In the last twenty years, the power output of the tokamaks has risen from a few watts to megawatts. If nature and fate are kind, these giant apparatuses will be the vanguard of reactors producing energy to spin the turbines of enough power plants to supply generous amounts of electricity for the entire world of our great-grandchildren.

During the last years of the twentieth century, some of the most important discoveries in fusion research were made with Princeton University's Tokamak Fusion Test Reactor (TFTR). This reactor achieved what was at the time a world record temperature of 510 million degrees Centigrade (for readers who like mind-boggling statistics, that is nearly 1 billion degrees Fahrenheit). The measurements of TFTR, precursor to bigger, better, and more expensive tokamaks, hint of the challenges that fusion research presents to our pocketbooks and patience. The outside dimensions of this doughnut were 24 feet in height and 38 feet in width. Its vacuum chamber weighed 80 tons, the magnetic field coils 587 tons. There was also a 15-ton titanium center column and a rugged stainless-steel support structure.

The expenses of fusion research, coupled with the appeal of its intellectual challenges, have melted away national insularity. In the

Workers inside the Tokamak Fusion Test Reactor vacuum chamber,
in Princeton, New Jersey.

late 1940s, American and Soviet fusion scientists began sharing at least
some of their research results, and in years of *glasnost* in the 1980s the
United States and the Soviet Union joined with the European Union
and Japan in a project to build the International Thermonuclear
Experimental Reactor (ITER). In the last decade of the twentieth
century, Princeton's TFTR generated 11 megawatts of power for a
third of a second: a world record. Impressive, but still a long way from
"burning" plasma. In the twenty-first century, ITER is supposed to
generate as many as 410 megawatts for up to 500 seconds. Such a performance would enable scientists to answer basic questions about
how to construct thermonuclear reactors not only to delight themselves and their colleagues, but to boil liquids or gases to spin turbines
to make electricity to keep our refrigerators cold and light our streets
at night. The ITER, if successful, will be to fusion power production
what Fermi's Chicago Pile was to fission power production.

The ITER will be awesome. It will have, for example, a single magnet weighing 925 tons. The costs will be in the billions, but the reward will be limitless—if it works. Since ITER's inception, Russia has replaced the Soviet Union in ITER, and South Korea and China have joined in. The United States resigned in 1998 because of the enormous expense and the possibility that a fusion reactor simply isn't practical. But in 2004, the U.S. government decided that the prospects of hydrogen fusion were not so dim after all and rejoined. The first ITER will be at Cadarache, France, starting in 2006. Construction will be difficult and long; experiments at the tokamak there won't start before 2014; and even if all goes well, the first practical hydrogen fusion power plants won't begin operation for some years after that.

When President George W. Bush announced the reentry of the United States into the ITER project, he stated that fusion energy would be commercially available within a few decades. Plasma physicists restrict themselves to more retractable predictions on the subject. For instance, consider the painfully constrained words of Richard Hazeltine, chairman of the U.S. Fusion Energy Sciences Advisory Committee, in 2002:

> It is not precipitous to emphasize an assumption that the fusion program, given continued scientific and technical accomplishment and sufficiently increased funding, could contribute to the US electric power grid within 35 years. It is not that 35 years is a best guess for timing fusion energy production, rather the assumption is that, with focused effort and appropriate support, the 35-year time scale becomes a credible goal.[11]

He can express himself more succinctly: "Fusion science is on the edge of vanishing—we need to go ahead and turn this damn thing on."[12]

[11]Richard Hazeltine to Ray Orbach, Sept. 13, 2002, www.ofes .FusionDocs.html. (Viewed on Oct. 12, 2003.)

[12]Geoff Brumfiel, "Just Around the Corner," *Nature* 436 (July 21, 2005), 318–20.

9

THE ANTHROPOCENE

No matter what else happens, this is the century in which we must learn to live without fossil fuels.

—*David Goodstein, physicist (2004)*[1]

It is very difficult to predict, especially the future.

—*Niels Bohr, physicist, Nobel laureate*[2]

Nobel laureate Paul Crutzen recommends that we drop the title, the Holocene, that geologists have given to the last ten thousand years, face up to hair-raising reality, and call it the Anthropocene—that is to say, the human epoch. He reasons that humans have gained so much power from fossil fuels that we have become a major factor, in some ways *the* major factor, in how the biosphere functions. We are as at least as powerful as ancient Nordic gods, from whom we haven't heard since the *Götterdämmerung.*

Our gain in power when we graduated from nearly exclusive dependence on recently derived sun energy—from muscles as the

---

[1]Goodstein, *Out of Gas: The End of the Oil Age,* 37.
[2]Richard L. Garwin and Georges Charpack, *Megawatts and Megatons: A Turning Point in the Nuclear Age* (New York: Alfred A. Knopf, 2001), 223. This aphorism was something Bohr repeated often.

prime mover—to buried sun energy—to coal, oil, and natural gas engines as prime movers, is almost too colossal to measure. John R. McNeill, an environmental historian, estimates that in the twentieth century humans used up a third more energy than they had in the hundred centuries between the dawn of agriculture and 1900. Jeffrey S. Dukes, an ecologist, estimates that between 1751 and 1998, we burned up 13,300 years worth of sunshine as manifested in plant and animal life here on our planet.[3] (The actual number of years must have been far, far greater than 13,300 because Dukes, in order to simplify his calculation, assumed a world 100 percent covered with plants and conditions totally conducive to their transformation to oil.)

The statements of McNeill and Dukes may strike the reader as too extreme to be believed. Let me provide specific illustrations of the difference between the amount of energy we could command before and after the fossil fuel or heat-engine revolution.

In 1586, nine hundred men and seventy-five horses directed by Dominico Fontana brought their strength to bear through thirty-seven windlasses to lift a 312-ton Egyptian obelisk from its base to move it from Curco di Nero to Piazza di San Pietro in Rome. Muscles are unlikely to achieve more than that because their strength per individual creature is slight and it is very difficult to focus the power of more than a few creatures on single specific tasks. Fontana's feat was as admirable an example of crowd control as of simple physics.

Rockets are our most efficient prime movers. On July 16, 1969, the Saturn V rocket, the biggest internal combustion engine yet, set off for the Moon with a roar so loud that one witness, Norman Mailer, thought that at last "man now had something with which to

[3]J. R. McNeill, *Something New Under the Sun: An Environmental History of the Twentieth-Century World* (New York: W. W. Norton & Company, 2000), 15; Jeffrey S. Dukes, "Burning Buried Sunshine: Human Consumption of Ancient Solar Energy," *Climatic Change*, 61 (November 2003), 37–38.

speak to God."[4] At launch the rocket weighed 6.4 million pounds. The first-stage engines burned nearly 5 million pounds of that sum in refined kerosene and liquid oxygen in 150 seconds to produce 160 million horsepower, or 7.5 million pounds of thrust. When the engines of the first stage stopped, the rocket was at an altitude of 41 miles traveling at 5,400 miles per hour. Getting from there to the Moon was relatively easy.

WE OF THE first years of the twenty-first century have access to more energy than we have the experience to wield intelligently. We of the richer societies make decisions we are not qualified to make almost every time we enter a voting booth or an automobile showroom or a grocery store, decisions that will in the long run have drastic effects on the lives of our descendants and on our planet. We of the poorer societies are too hungry in the short run for a decent standard of living to make wise decisions about the long run. As one citizen of Brazil put it, "We're not going to stay poor because the rest of the world wants to breathe."[5]

A good first step toward making informed decisions on the long run might be to decide what *normal* means to us *vis-à-vis* our energy situation because that is the baseline from which we launch our inquiries. "Normal" is often taken to mean how things are now. For the citizens of the rich societies today, normal involves vast expenditures of energy to empower a multitude of devices from aircraft carriers to desktop computers. This definition has penetrated deeply into the attitudes of people everywhere. For instance, Iraqis, citizens of a war-wracked nation in 2005, complain bitterly because the supply of the miracle juice, electricity, is not continuous in their cities.

But their view—our view—of what is normal is wrong. Street-

---

[4]Norman Mailer, *Of a Fire on the Moon* (Boston: Little, Brown, 1970), 100.

[5]Michael Williams, *Deforesting the Earth: From Prehistory to Global Crisis* (Chicago: University of Chicago Press, 2002), 499.

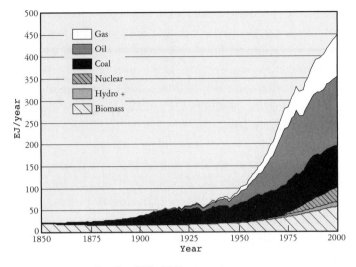

Graph of World Energy, 1850–2000

Human energy consumption as measured in exajoules. A joule is the power required to produce one watt for one second; an exajoule is a million million million joules.

lights, toasters, automobiles and other such devices which are found or at least coveted everywhere are artifacts of a fossil fuel and prime mover revolution that in its full florescence is no more than a century and a half old (see the graph above). Since that is longer than living memory, most of us in the richer societies can only recall times of immediate access to abundant energy. That abundance tempts us, successfully, to believe, for instance, that having energy flow down lines from far away and illuminate our rooms when we flip the switch is normal rather than miraculous.

The most obvious challenge to our capacity to extend such a miracle to all of humanity is demographic. In spite of its hideous wars, pandemics, disruptions of family structure, and industrial contaminations, the Anthropocene has witnessed a population increase in every major region of the world. During the last forty years, the number of people doubled on earth. We are the first humans to witness such an explosion in such a short time.

It is likely, barring catastrophes such as comet strikes, an accelerated AIDS pandemic, or thermonuclear exchanges, that the population increase will continue for at least a generation or two more. Fertility rates have begun to decline and the rate of increase is slowing, but even so it is probable that between 2000 and 2050 we will add more people than were alive in 1950. Most of the new arrivals will be born in the poorer nations, where the fertility rates are dropping more slowly than in the developed world. Most of the new arrivals will live in cities, none of which are self-sustaining. In 2000, only four of the nineteen cities with populations of 10 million or more were in the rich nations.

By 2050, the citizenry of the rich societies will include large numbers of the middle-aged and elderly, who in many cases will be, speaking broadly, consuming more than they will be currently producing. In the poorer societies there will be enormous numbers of people in what are potentially the most productive years of life, but who, in order to produce, will need, food, clothing, education, and jobs in unprecedented quantities.

The demand for more power in the twenty-first century will be enormously greater than in the twentieth. At present rates, the demand will double by 2035. We will be hard put to procure enough fossil fuel at prices low enough to meet that demand, and in any event the health of the biosphere cannot tolerate the amounts of pollution and the rising temperatures that would come with multiplications of fossil fuel combustion as currently practiced.

WE ARE, I'M glad to say, not totally unprepared for the energy crisis that is coming. We know how to raise fossil fuel engines' efficiency, to minimize their emissions, and so to stretch out the time we have to learn our new role as, we hope, benign gods of the biosphere. We have already accumulated a lot of experience in building and maintaining windmills, photovoltaic cells, hydrogen fuel cells, and so on. New and truly safe nuclear power plants may well be possible. Surely we can bury radioactive wastes so deep that they will not endanger us or our descendants. And perhaps our physicists will turn another of their glorious tricks and successfully domesticate hydrogen fusion to initiate a golden age.

We have reason to believe that we are capable of environmental sanity; but first we have to accept that the way we live now is new, abnormal, and unsustainable. Very few of us would choose to reject the benefits of coal, oil, and natural gas and return to the good old days of no shoes, hunger pangs, and chills. But the truth remains that winning streaks are rarely permanent.

Fossil fuels in some ways remind me of amphetamines ("speed" in the vernacular). Amphetamine pills, like the fossil fuels, are enormously stimulating. The taker doesn't tire, seems to think and in general to function better and faster, and is happier. But people who take the pills can become dependent on them, which is unfortunate because their supply cannot be guaranteed. Furthermore, people who take them may suffer distorted comprehensions of reality. Lastly, amphetamines, taken in large dosages for too long, are poisonous.

We children of the sun may be standing on the peak of our energy achievements poised for the next quantum leap upward—like Paleolithic hunter-gatherers scratching wolf puppies behind the ears or like Newcomen and Watt noticing that steam has push as well as heat. Or we may be teetering there, destined to participate in nature's standard operational procedure of pairing a population explosion with a population crash.

CODA: ESTRANGEMENT FROM THE SUN

The electricity blackout of August 2003 along the eastern region was the biggest in the history of North America. Various officials, harassed for quick explanations, blamed it on lightning, a nuclear power accident, and the "blaster" computer worm. Somebody who said he was an Islamic terrorist claimed credit for his band of brothers. The truth was that on the fourteenth day of the month in Ohio, the Hanna-Juniper electric transmission line sagged into tree branches long overdue for pruning and a short circuit occurred, challenging an electrical grid system already wavering under the burden of normal demand.

This initiated an epidemic of disconnections between power plants, most of them driven by fossil fuel combustion or falling water (both vehicles of sun energy), and the consumers of the electricity. A hundred plants, linked together for purposes of reciprocal assistance, failed. Like a line of staggering drunks, elbows locked together for mutual support, they tumbled down en masse.

Fifty million children of the sun living in 9,300 square miles of the northeastern United States and adjoining Canadian provinces, the homelands of some of the wealthiest and most technologically sophisticated people in the world, abruptly found themselves without electricity.

Elevators stopped, trapping passengers. Computer screens went blank. Automatic bank machines reverted to being uncommunicative blocks of metal, plastic, and glass. Wire telephone service continued, but couldn't keep up with the demand. Cellular phones, being battery-powered, continued to work, but the cellular phone centers, where the antennas that forward the calls are located, failed. Airline traffic stumbled as means of communication from ground to aircraft faltered and the equipment for checking electronic tickets turned off.

In several cities, Cleveland, for instance, failure of electric pumps endangered the purity of the drinking water and citizens were advised to boil their water—a dubious suggestion if you had an electric stove. Some urban sewage plants stopped functioning: New York City's dumped 490 million gallons of raw waste into its waterways. The weather was hot and air conditioners were inoperative. Mayor Michael Bloomberg advised New Yorkers "to go home, open up your windows, drink lots of liquids." He told them they would be doing a lot of walking.

True enough: Traffic lights blinked out and traffic jams paralyzed above-ground urban transportation. Railroad schedules crumbled into chaos. Subway trains rolled to a halt; it took two and a half hours in New York to guide stranded riders out of the tunnels and back to street level. Getting transportation out of Manhattan back home to the outer boroughs and suburbs was nearly impossible and some daytime Manhattanites slept in the park or on city streets that night. Thousands of people began to walk out of Manhattan over the bridges, the only practical means of mass exit left.

There was, amazingly, very little looting or mugging in the darkness. Some urbanites got to see the Milky Way for the first time in their lives, a pleasant compensation for inconvenience. There were a few unintentional fires started by people inexperienced in the use of candles.

THE BLACKOUT PARALYZED society for no more than hours in most places and for never more than a day or so elsewhere. For many people, the brief loss of energy being delivered from the Sun as electricity turned out to be more of an adventure than a disaster. For others, the blackout of 2003 was a premonitory vision, acutely foreshortened to enhance comprehension, of our possible course in the next few centuries.

BIBLIOGRAPHY

PART I: THE LARGESS OF THE SUN

Garcilaso de la Vega, El Inca, *Royal Commentaries of the Incas and General History of Peru*, trans. Harold V. Livermore. Austin: University of Texas Press, 1966.

Gribbin, John. *Almost Everyone's Guide to Science: The Universe, Life, and Everything*. New Haven: Yale University Press, 1999.

Lang, Kenneth R. *The Cambridge Encyclopedia of the Sun*. Cambridge: Cambridge University Press, 2001.

Smil, Vaclav. *The Earth's Biosphere: Evolution, Dynamics, and Change*. Cambridge, MA: MIT Press, 2002.

Vogel, Steven. *Prime Mover, A Natural History of Muscle*. New York: W. W. Norton & Company, 2001.

Zirker, Jack B. *Journey from the Center of the Sun*. Princeton: Princeton University Press, 2002.

CHAPTER 1: FIRE AND COOKING

Abrams, H. Leon. "Fire and Cooking as a Major Influence on Human Cultural Advancement: An Anthropological Perspective," *Journal of Applied Nutrition*, 36 (1986), 24–31.

———. "Hominid Proclivity for Sweetness: An Anthropological View," *Journal of Applied Nutrition,* 39 (1987), 35–41.

Aiello, Leslie C., and Peter Wheeler, "The Expensive-Tissue Hypothesis: The Brain and the Digestive System in Human and Primate Evolution," *Current Anthropology*, 36 (1995), 199–221.

Byrne, Richard. *The Thinking Ape: Evolutionary Origins of Intelligence*. Oxford: Oxford University Press, 1996.

Clark, J. D., and J. W. K. Harris. "Fire and Its Roles in Early Hominid Lifeways," *African Archaeological Review*, 3 (1985), 3–27.

Darwin, Charles. *The Voyage of the Beagle.* New York: Doubleday & Co., 1962.

———. *The Origin of Species by Means of Natural Selection and The Descent of Man and Selection in Relation to Sex.* New York: Modern Library, n.d.

Goudsblom, J. "The Human Monopoly on the Use of Fire: Its Origins and Conditions," *Human Evolution*, 1 (1986), 517–23.

Hladik, C. W., D. J. Chivers, and P. Pasquet. "On Diet and Gut Size in Non-human Primates and Humans: Is There a Relation to Brain Size?" *Current Anthropology*, 40 (1999), 695–97.

Klein, Richard G. *The Human Career: Human Biological and Cultural Origins.* Chicago: University of Chicago Press, 1989.

Lambert, Craig. "The Way We Eat Now," *Harvard Magazine*, 106 (May–June 2004), 50–58, 98–99.

Leonard, William. "We Humans Are Strange Primates," *Scientific American*, 287 (December 2002), 106–15.

Lewin, Roger. *Human Evolution: A Core Textbook.* Malden, MA: Blackwell Science, 1998.

Milton, Katherine. "Diet and Primate Evolution," *Scientific American*, 269 (August 1993), 86–93.

Ofek, Haim. *Second Nature: Economic Origins of Human Evolution.* Cambridge: Cambridge University Press, 2001.

Pennisi, Elizabeth. "Did Cooked Tubers Spur the Evolution of Big Brains?" *Science*, 283 (March 26, 1999), 2004–05.

Smith, Steven R. "Hominid Use of Fire in the Lower and Middle Pleistocene," *Current Anthropology*, 30 (1989), 1–26.

Stahl, Ann Brower. "Hominid Dietary Selection Before Fire," *Current Anthropology*, 25 (1984), 151–68.

Stanford, Craig B. "The Ape's Gift: Meat-eating, Meat-sharing, and Human Evolution," in Frans B. M. de Waal, ed., *Tree of Origin: What Primate Behavior Can Tell Us About Human Social Evolution.* Cambridge, MA: Harvard University Press, 2001, 95–119.

———. *The Hunting Apes: Meat Eating and the Origins of Human Behavior.* Princeton: Princeton University Press, 1999.

———. *Significant Others: The Ape-Human Continuum and the Quest for Human Nature.* New York: Basic Books, 2001.

Tattersall, Ian. *Becoming Human: Evolution and Human Uniqueness.* New York: Harcourt Brace, 1998.

Wrangham, Richard. "The Raw and the Stolen," *Current Anthropology*, 40 (1999), 567–620.

———. "The Evolution of Cooking; A Talk with Richard Wrangham." Edge Foundation, 2001, www.edge.org. (Viewed Dec. 11, 2003.)

———. "Out of the Pan, into the Fire: How Our Ancestors' Evolution Depended on What They Ate," in de Waal, ed., *Tree of Origin: What Primate Behavior Can Tell Us About Human Social Evolution.*

"Yamanas," tierradelfuego.org.ar/museo/ushuaia/indigenas/indigenaso2-l .htm. (Viewed Dec. 11, 2003.)

CHAPTER 2: AGRICULTURE

Budiansky, Stephen. *The Covenant of the Wild: Why Animals Chose Domestication.* Leesburg, VA: Terrapin Press, 1995.

———. *The Truth About Dogs.* New York: Viking/Penguin, 2001.

Cambridge World History of Food, ed. Kenneth F. Kiple and Kriemhild Cornée Ornelas. 2 vols. Cambridge: Cambridge University Press, 2000.

Chance, Norman. "The 'Lost' Reindeer of Arctic Alaska," http:// arcticcircle.uconn.edu/NatResources/reindeer.html. (Viewed May 28, 2004.)

Clutton-Brock, Juliet. *Horse Power: A History of the Horse and the Donkey in Human Societies.* Cambridge: Harvard University Press, 1992.

———. *A Natural History of Domesticated Mammals.* Cambridge: Cambridge University Press, 1999.

Cohen, Mark Nathan. *The Food Crisis in Prehistory: Overpopulation and the Origins of Agriculture.* New Haven: Yale University Press, 1977.

Diamond, Jared. "The Worst Mistake in the History of the Human Race," *Discover*, 8 (May 1987), 64–66.

———. "Evolution, Consequences and Future of Plant and Animal Domestications," *Nature*, 418 (Aug. 8, 2002), 700–07.

Elton, Charles S. *The Ecology of Invasions by Animals and Plants.* London: Methuen & Co., 1958.

Fagan, Brian. *The Long Summer: How Climate Changed Civilization.* New York: Basic Books, 2004.

Hadwen, Seymour, and Lawrence J. Palmer. "Reindeer in Alaska," http://

www.alaskool.org/projects/reindeer/history/usda1922/AKRNDEER
.html#top. (Viewed May 28, 2004.)

Hare, Brian, et al. "The Domestication of Social Cognition in Dogs," *Science*, 298 (Nov. 22, 2002), 1634–36.

"Jackson, Sheldon," *Dictionary of American Biography,* ed. Dumas Malone. Vol. 9. New York: Charles Scribner's Sons, 1922.

Lang. *The Cambridge Encyclopedia of the Sun.*

Larsen, C. S., and Alfred W. Crosby. "A Biohistory of Health and Behavior in the Georgia Bight: The Agricultural Transition and the Impact of European Contact," in Richard Steckel and Jerome Rose, eds., *The Backbone of History: Health and Nutrition in the Western Hemisphere.* Cambridge: Cambridge University Press, 2002, 406–39.

Lazell, J. Arthur. *Alaskan Apostle: The Life Story of Sheldon Jackson.* New York: Harper & Brothers, 1960.

Leonard, Jennifer A., et al. "Ancient DNA Evidence for Old World Origins of New World Dogs," *Science*, 298 (Nov. 22, 2002), 1613–16.

Mears, John A. "Agricultural Origins in Global Perspective," in Michael Adas, ed., *Agricultural and Pastoral Societies in Ancient and Classical History.* Philadelphia: Temple University Press, 2001, 36–70.

Miklosi, Ádám, et al. "A Simple Reason for a Big Difference: Wolves Do Not Look at Humans, But Dogs Do," *Current Biology*, 13 (April 29, 2003), 763–66.

Perkins, John H. *Geopolitics and the Green Revolution: Wheat, Genes, and the Cold War.* New York: Oxford University Press, 1997.

Pringle, Heather. "The Slow Birth of Agriculture," *Science*, 282 (Nov. 20, 1998), 1446–56.

Savolainen, P., et al. "Genetic Evidence for an East Asian Origin of Domestic Dogs," *Science*, 298 (Nov. 22, 2002), 1610–13.

Schaechter, Elio. *In the Company of Mushrooms: A Biologist's Tale.* Cambridge, MA: Harvard University Press, 1977.

Shuttlesworth, Dorothy. "Ant," *Encyclopedia Americana*, Vol. 2. Danbury, CT: Grolier, 1998, 20–26.

CHAPTER 3: THE COLUMBIAN EXCHANGE

Collier, Peter, and David Horowitz. *The Fords, An American Epic.* New York: Summit Books, 1987.

Cambridge World History of Food, ed. Kiple and Ornelas.

Crosby, Alfred W. *Germs, Seeds, and Animals: Studies in Ecological History.* Armonk, NY: M. E. Sharpe, 1994.

———. *The Columbian Exchange: Biological and Cultural Consequences of 1492.* 2nd ed. Westport, CT: Praeger, 2003.

———. *Ecological Imperialism: The Biological Expansion of Europe, 900–1900.* 2nd ed. Cambridge: Cambridge University Press, 2004.

Kinealy, Christine. *The Great Calamity: The Irish Famine, 1845–52.* Boulder, CO: Roberts Rinehart, 1994.

Mazumdar, Sucheta. "The Impact of New World Food Crops on the Diet and Economy of China and India, 1600–1900," in Raymond Grew, ed., *Food in Global History.* Boulder, CO: Westview, 1999.

McNeill, William H. "American Food Crops in the Old World," in Herman J. Viola and Carolyn Margolis. eds., *Seeds of Change: A Quincentennial Commemoration.* Washington, DC: Smithsonian Press, 1991.

Salaman, Redcliffe. *The History and Social Influence of the Potato.* Rev. ed. Cambridge: Cambridge University Press, 1985.

Woodham-Smith, Cecil. *The Great Hunger: Ireland 1845–49.* New York: New American Library, 1964.

PART II: FOSSILIZED SUNSHINE

Braudel, Fernand. *Civilization and Capitalism, 15th–18th Century.* Vol. 1: *The Structures of Everyday Life: The Limits of the Possible*, trans. Siân Reynolds. New York: Harper & Row, 1981.

Cleal, Christopher J., and Barry A. Thomas. *Plant Fossils: The History of Land Vegetation*, Woodbridge, UK: Boydell Press, 1999.

Dukes, Jeffrey S. "Burning Buried Sunshine: Human Consumption of Ancient Solar Energy," *Climatic Change*, 61 (November 2003), 31–44.

Gimpel, Jean. *The Medieval Machine: The Industrial Revolution of the Middle Ages.* New York: Henry Holt & Co., 1976.

Goodstein, David. *Out of Gas: The End of the Age of Oil.* New York: W. W. Norton & Company, 2004.

Pacey, Arnold. *Technology in World Civilization: A Thousand-Year History.* Cambridge, MA: MIT Press, 1990.

Smil, Vaclav. *Energy in World History.* Boulder, CO: Westview Press, 1994.

Taylor, Thomas N., and Edith L. Taylor. *The Biology and Evolution of Fossil Plants.* Englewood Cliffs, NJ: Prentice Hall, 1993.

Vogel, Steven. "A Short History of Muscle-Powered Machines," *Natural History*, III (March 2002), 85–91.

CHAPTER 4: COAL AND STEAM

Bly, Nellie. *Around the World in Seventy-two Days.* New York: Pictorial Weeklies and Brentanos, 1890. http://erc.lib.umn.edu:80/dynaweb/travel/blyaroun@Generic__BookTextView/201;pt/201. (Viewed March 9, 2004.)

Braudel, Fernand. *Civilization and Capitalism, 15th–18th Century.* Vol. 1.

———. *Civilization and Capitalism, 15th–18th Century.* Vol. 3: *The Perspective of the World,* trans. Siân Reynolds. New York: Harper & Row, 1984.

———. *The Identity of France.* Vol 2: *People and Production,* trans. Siân Reynolds. New York: HarperCollins, 1990.

Cleal and Thomas. *Plant Fossils: The History of Land Vegetation.*

Crosby, Alfred W. *Throwing Fire: Projectile Technology Through History.* Cambridge: Cambridge University Press, 2002.

Dickinson, H. W. "The Steam Engine to 1830," in Charles Singer et al., eds., *A History of Technology.* Vol. 4: *The Industrial Revolution, c. 1750–c. 1850.* London: Oxford University Press, 1958.

Engels, Friedrich. *The Condition of the Working Class in England.* New York: Viking/Penguin, 1987.

Freese, Barbara. *Coal: A Human History.* Cambridge, MA: Perseus Publishing, 2003.

Gimpel. *The Medieval Machine: The Industrial Revolution of the Middle Ages.*

Kroeger, Brooke. *Nellie Bly: Daredevil, Reporter, Feminist.* New York: Random House, 1994.

Landes, David S. *The Wealth and Poverty of Nations: Why Some Are So Rich and Some So Poor.* New York: W. W. Norton & Company, 1998.

McNeill, J. R., and William McNeill. *The Human Web: A Bird's Eye View of World History.* New York: W. W. Norton & Company, 2003.

Marks, Jason. *Around the World in 72 Days.* New York: Gemittarius Press, 1993.

Marks, Robert B. *The Origins of the Modern World: A Global and Ecological Narrative.* Lanham, MD: Rowman & Littlefield, 2002.

Needham, Joseph. *The Shorter Science and Civilisation in China,* abridged

Colin A. Ronan. 5 vols. Cambridge: Cambridge University Press, 1978–95.

Pacey, Arnold. *Technology in World Civilization: A Thousand-Year History.* Cambridge, MA: MIT Press, 1990.

Pomeranz, Kenneth. *The Great Divergence: China, Europe, and the Making of the Modern World Economy.* Princeton: Princeton University Press, 2000.

Richards, John F. *The Unending Frontier: An Environmental History of the Early Modern World.* Berkeley: University of California Press, 2003.

Smil, *Energy in World History.*

Taylor and Taylor. *The Biology and Evolution of Fossil Plants.*

Thurston, Robert H. *The History of the Growth of the Steam Engine.* New York: D. Appleton & Co., 1878. (There are later editions, the most recent in 1972.)

Vries, Jan de, and Ad van der Woude. *The First Modern Society: Success, Failure, and Perseverance of the Dutch Economy, 1500–1815.* Cambridge: Cambridge University Press, 1997.

The Works of Daniel Webster. Vol. 1. Boston: Charles C. Little & James Brown, 1851.

Williams, Michael. *Deforesting the Earth from Prehistory to Global Crisis.* Chicago: University of Chicago Press, 2003.

Zeeuw, J. W. de. "Peat and the Dutch Golden Age," *A.A.G. Bijdagen,* 21 (1978), 3–31.

CHAPTER 5: OIL AND THE ICE

Adams, Henry. *The Education of Henry Adams.* New York: Penguin Books, 1995.

Blond, Georges. *The Marne: The Battle That Saved Paris and Changed the Course of the First World War,* trans. H. Easton Hart. London: Prion Books, 2002.

Black, Brian. *Petrolia: The Landscape of America's First Oil Boom.* Baltimore: Johns Hopkins University Press, 2000.

Bryant, Lynwood. "The Silent Otto," *Technology and Culture,* 7 (Spring 1966), 184–200.

———. "The Origin of the Four-Stroke Cycle," *Technology and Culture,* 8 (April 1967), 178–98.

Byers, Edward. *The Nation of Nantucket: Society and Politics in an Early*

American Commercial Center, 1660–1820. Boston: Northeastern University Press, 1987.

George, Dorothy N. *London Life in the Eighteenth Century*. New York: Harper & Row, 1964.

Inwood, Stephen. *A History of London*. New York: Carroll & Graf, 1998.

Lichty, Lester C. *Combustion Engine Processes*. New York: McGraw-Hill, 1967.

Matthews, L. Harrison. "A Note on Whaling," in Charles Singer et al., eds., *History of Technology*, Vol. 4: *The Industrial Revolution, c. 1750–c.1850*. London: Oxford University Press, 1958, 55–63.

New Encyclopedia of Motorcars, 1885 to Present, ed. G. N. Georgano. New York: E. P. Dutton & Co., 1982.

Schivelbusch, Wolfgang. *Disenchanted Night: The Industrialization of Light in the Nineteenth Century*, trans. Angela Davies. Berkeley: University of California Press, 1988.

Stackpole, Edouard A. *Whales and Destiny: The Rivalry Between America, France, and Britain for Control of the Southern Whale Fishery, 1785–1825*. Amherst, MA: University of Massachusetts Press, 1972 .

Starbuck, Alexander. *History of the American Whale Fishery*. Secaucus, NJ: Castle Books, 1989 (1878).

Wolfe, John J. *Brandy, Balloons, and Lamps: Ami Argand, 1750–1803*. Carbondale, IL: Southern Illinois Press, 1999.

Yergin, Daniel. *The Prize: The Epic Quest for Oil, Money, and Power*. New York: Simon & Schuster, 1992.

CHAPTER 6: ELECTRICITY

Henry Adams: Novels, Mont Saint Michel, The Education. New York: Literary Classics of the United States, 1984, 106.

Burg, David F. *Chicago's White City of 1893*. Lexington, KY: University Press of Kentucky, 1976.

ColumbianExposition. http://www.chicagohs.org/history/expo/ex2html. (Viewed April 30, 2004.)

Elton, Arthur. "Gas for Light and Heat," in Singer et al., eds., *A History of Technology*, Vol. 4, 64–98.

Encyclopedia of the United States in the Nineteenth Century, ed. Paul Finkelman. New York: Charles Scribner's Sons, 2001.

Halstead, Murat. "Electricity at the Fair," *Cosmopolitan*, 15 (September

1893). http://www.boondocksnet.com/expos/wfe_1893_cosmo_electricity
.html. (Viewed April 14, 2004.)

Jarvis, C. Mackechnie. "The Distribution and Utilization of Electricity,"
in Charles Singer et al., eds., *A History of Technology.* Vol. 5: *The Industrial Revolution, c. 1750–c.1850.* London: Oxford University Press, 1958,
208–34.

Jones, Lois Stodieck. *The Ferris Wheel.* Reno, NE: Grace Dangberg Foundation, 1984.

Jonnes, Jill. *Empires of Light: Edison, Tesla, Westinghouse, and the Race to
Electrify the World.* New York: Random House, 2003.

Lenin, V. I. "Report to the Eighth Congress of the All-Russia Congress of
Soviets," http://www.marx.org/archive/lenin/works/1920/8thcong/ch03
.htm#35n. (Viewed Aug. 24, 2004.)

Meyer, Herbert W. *A History of Electricity and Magnetism.* Norwalk, CT:
Burndy Library, 1972.

Nye, David E. *Electrifying America: Social Meanings of a New Technology,
1880–1940.* Cambridge, MA: MIT Press, 1990.

Schivelbusch, Wolfgang. *Disenchanted Night: The Industrialization of Light
in the Nineteenth Century,* trans. Angela Davies. Berkeley: University of
California Press, 1988.

Shectman, Jonathan. *Groundbreaking Scientific Experiments, Inventions
and Discoveries of the 18th Century.* Westport, CT: Greenwood Press,
2003.

PART III: ENERGY AT THE TURN OF THE
THIRD MILLENNIUM

Berenson, Alex. "An Oil Enigma: Production Falls Even as Reserves Rise,"
New York Times, June 12, 2004.

Berger, John J. *Charging Ahead: The Business of Renewable Energy and
What It Means for America.* Berkeley: University of California Press,
1997.

Berinstein, Paula. *Alternative Energy: Facts, Statistics, and Issues.* Westport,
CT: Oryx Press, 2001.

Cohen, Joel E. *How Many People Can the Earth Support?* New York:
W. W. Norton & Company, 1995.

Deffeyes, Kenneth S. *Hubbert's Peak: The Impending World Oil Shortage.* Princeton: Princeton University Press, 2001.

Encyclopedia of World Environmental History, ed. Shepard Krech III, J. R. McNeill, and Carolyn Merchant. 3 vols. New York: Routledge, 2004.

Gerth, Jeff, and Stephen Labaton, "Oman's Oil Yield in Decline, Shell Data Shows," *New York Times*, April 8, 2004.

Grant, Paul M. "Hydrogen Lifts Off—with a Heavy Load," *Nature*, 424 (July 10, 2003), 129–30.

Hall, Charles, et al., "Hydrocarbons and the Evolution of Human Culture," *Nature*, 426 (Nov. 20, 2003), 318–22.

Hoffmann, Peter. *Tomorrow's Energy: Hydrogen, Fuel Cells, and the Prospects for a Cleaner Planet.* Cambridge, MA: MIT Press, 2002.

Holdren, John P. "The Energy Climate Challenge: Issues for the New U.S. Administration," *Environment* (June 2001).

Johansson, Thomas B., et al. *Renewable Energy: Sources for Fuel and Electricity.* Washington DC: Island Press, 1993.

Kirsner, Scott. "Wind Power's New Current," *New York Times*, Aug. 28, 2003.

Loftness, Robert L. "Power, Electric," *Encyclopedia Americana.* Vol. 22. Danbury, CT: Scholastic Library Publishing, 2003, 501–08.

McNeill, J. R. *Something New Under the Sun: An Environmental History of the Twentieth-Century World.* New York: W. W. Norton & Company, 2001.

Smil, Vaclav, *Energy in World History.*

———. *Energy at the Crossroads: Global Perspectives and Uncertainties.* Cambridge, MA: MIT Press, 2003.

Talwani, Manik. "Will Calgary Be the Next Kuwait?" *New York Times*, Aug. 14, 2003, Op-Ed page.

Vaitheeswaran, Vijay V. *Power to the People.* New York: Farrar, Straus & Giroux, 2003.

Williams, Michael. *Deforesting the Earth: From Prehistory to Global Crisis.* Chicago: University of Chicago Press, 2003.

CHAPTER 7: FISSION

Ahman, Birgitta. "Transfer of Radiocaesium via Reindeer Meat to Man— Effects of Countermeasures Applied in Sweden Following the Chernobyl Accident," *Journal of Environmental Radioactivity,* 46 (1999), 113–20.

Atkins, Stephen E. *Historical Encyclopedia of Atomic Energy.* Westport, CT: Greenwood Press, 2000.

Beach, Hugh. "Perceptions of Risk, Dilemmas of Policy: Nuclear Fallout in Swedish Lapland," *Social Science and Medicine*, vol. 30, no. 6 (1990), 729–38.

———. Personal communications.

Broadbent, N. D. "Chernobyl Radionuclide Contamination and Reindeer Herding in Sweden," *Colloquium Antropologia*, Vol. 2. Zagreb, Yugoslavia, 1986, 231–42.

Butler, Declan. "Nuclear Power's New Dawn," *Nature*, 429 (May 20, 2004), 238–40.

Cassedy, Edward S. *Prospects for Sustainable Energy, A Critical Assessment.* Cambridge: Cambridge University Press, 2000.

———, and Peter Z. Grossman. *Introduction to Energy:Resources, Technology, and Society.* Cambridge: Cambridge University Press, 1998.

"Chernobyl Nuclear Disaster." http//:www.chernobyl.co.uk.environmental .html.

The First Reactor, U.S. Department of Energy/NE-0046 (December 1982).

Garwin, Richard L., and Georges Charpak. *Megawatts and Megatons: A Turning Point in the Nuclear Age?* New York: Alfred A. Knopf, 2001.

Hecht, Gabrielle, *The Radiance of France: Nuclear Power and National Identity After World War II.* Cambridge, MA: MIT Press, 1998.

Knight, Robin. "The Legacy of Chernobyl: Disaster for the Lapps," *U.S. News & World Report*, 102 (March 23, 1987), 36.

Kragh, Helge. *Quantum Generations: A History of Physics in the Twentieth Century.* Princeton: Princeton University Press, 1999.

Medvedev, Grigori. *The Truth About Chernobyl,* trans. Evelyn Rossitor. New York: Basic Books, 1991.

Nuclear Energy Institute, "The Chernobyl Accident and Its Consequences," July 2000. http://www.nei.org/doc.asp?docid=456. (Viewed Aug. 31, 2004).

The Nuclear History Site. http://nuclearhistory.tripod.com/history.html#top.

Osif, Bonnie, Anthony J. Baratta, and Thomas W. Conkling. *TMI 25 Years Later: The Three Mile Island Nuclear Power Plant Accident and Its Impact.* University Park, PA; Pennsylvania University Press, 2004.

Rhodes, Richard. *Dark Sun: The Making of the Hydrogen Bomb.* New York: Simon and Schuster, 1995.

Savchenko, V. K. *The Ecology of the Chernobyl Catastrophe.* Paris: UNESCO, 1995.

Stamstag, Tony. "Norway's Radioactive Reindeer," BBC World News, Europe, Dec. 24, 2000.

Vargo, G. J. *The Chernobyl Accident, Radiation, and Health Concerns.* Richland, WA: Pacific Northwest National Laboratory, 2000.

Wald, Matthew L. "Court Sets Back Federal Project on Atomic Waste Site's Safety," *New York Times*, July 10, 2004.

———. "Ruling on Nuclear Site Leaves Next Move to Congress," *New York Times*, July 15, 2004.

World Spaceflight News. *21st Century Complete Guide to Nuclear Power: Encyclopedic Coverage of Power Plants, Reactors, Fuel Processing, NRC and Department of Energy Regulations, Radioactive Waste, New Plant Designs.* Progressive Management, 2002.

CHAPTER 8: FUSION

"Bringing a Star to Earth." U.S. Department of Energy. www.ofes.fusion.doe.gov/FusionDocs, html. (Viewed Nov. 7, 2003.)

Brumfiel, Geoff. "Just Around the Corner," *Nature* 436 (July 21, 2005), 318–20.

Cassedy. *Prospects for Sustainable Energy, A Critical Assessment.*

Chang, Kenneth. "U.S. Plans to Rejoin Project to Develop Fusion Reactor," *New York Times*, Jan. 31, 2003.

———. "New Fusion Method Offers Hope of New Energy Source," *New York Times*, April 8, 2003.

Close, Frank. *Too Hot to Handle: The Race for Cold Fusion.* Princeton: Princeton University Press, 1991.

Conn, Robert W., "The Engineering of Magnetic Fusion Reactors," *Scientific American*, 249 (October 1983), 60–71.

Davis, Bennett. "Reasonable Doubt," *New Scientist*, 177 (Mar. 29, 2003), 36–38.

Garwin and Charpak, *Megawatts and Megatons: Turning Point in the Nuclear Age?*

Huizenga, John R. *Cold Fusion: The Scientific Fiasco of the Century.* Rochester, NY: University of Rochester Press, 1992.

Kragh. *Quantum Generations: A History of Physics in the Twentieth Century.*

Mallove, Eugene F. *Fire from Ice: Searching for the Truth Behind the Cold Fusion Furor.* New York: John Wiley & Sons, 1991.

National Academies, National Research Council, Letter Report, Burning

Plasma Assessment, Dec. 20, 2002. http://books.nap.edu/books/N1000487/html/1.html. (Viewed Nov. 6, 2003.)

"150-ton Magnet Pulls World Closer to Nuclear Fusion as a Potential Source of Energy." Massachusetts Institute of Technology, September 23, 2002. BrightSurf.com. (Viewed Nov. 7, 2003.)

Putterman, S. J. "Sonoluminescence: Sound into Light," *Scientific American*, 272 (February 1995), 46–51.

Sakharov, Andrei. *Memoirs*, trans. Richard Lourie. New York: Alfred A. Knopf, 1991.

Scheider, Walter. *A Serious, But Not Ponderous Book About Nuclear Energy.* Ann Arbor, MI: Cavendish Press, 2001.

Sheffield, John. *Fusion Energy in India's Long-Term Future.* Knoxville, TN: Joint Institute for Energy and Environment Report Number JIEE 2003-03, May 2003.

Smith, Craig S. "France Will Get Fusion Reactor to Seek a Furture Energy Source," *New York Times* (June 29, 2005).

Taleyarkhan, R. P., et al. "Evidence for Nuclear Emissions During Acoustic Cavitation," *Science*, 295 (March 2002), 1868–73.

Weiss, P. "Bubble Fusion," *Science News*, 165 (Mar. 6, 2004), 149.

Wells, H. G. *The World Set Free: A Story of Mankind.* New York: E. P. Dutton & Co., 1914, 37–38.

CHAPTER 9: THE ANTHROPOCENE

Butler. "Nuclear Power's New Dawn," *Nature*, 238–40.

CNN.com, "Major Power Outage Hits New York, Other Large Cities," http://www.cnn.com/2003/US/08/14/power.outage. (Viewed Oct. 30, 2004.)

CNN.com, "Power Returns to Most Areas Hit by Blackout," http://www.cnn.com/2003/US/08/15/power.outage. (Viewed Oct. 30, 2004.)

Cohen, Joel E. *How Many People Can the Earth Support?* New York: W. W. Norton & Co., 1995.

———. "Human Population: The Next Half Century," *Science*, 302 (Nov. 14, 2003), 1172–75.

Crutzen, Paul. "Introducing the Anthropocene," *International Geosphere-Biosphere Programme Newsletter* (May 2000).

Dukes, Jeffrey S., "Burning Buried Sunshine: Human Consumption of Ancient Solar Energy," *Climatic Change*, 61 (November 2003), 31–44.

Ehrlich, Paul, and Anne Ehrlich. *One with Nineveh: Politics, Consumption, and the Human Future.* Washington, DC: Island Press, 2004.

Goodstein. *Out of Gas: The End of the Oil Age.*

Holdren, John P. "Environmental Change and the Human Condition," *American Academy of Arts & Sciences Bulletin,* 57 (Fall 2003), 24–31.

"Introducing the Anthropocene," *Geology Newsletter.* http://geology.about .com/library/weekly/aso80402a.htm. (Viewed May 10, 2004.)

Mailer, Norman. *Of a Fire on the Moon.* Boston: Little, Brown, 1970.

McNeill, J. R. *Something New Under the Sun: An Environmental History of the Twentieth-Century World.* New York: W. W. Norton & Company, 2000.

Roberts, Paul. *The End of Oil: On the Edge of a Perilous New World.* Boston: Houghton Mifflin, 2003.

Smil, Vaclav. *Energy in World History.*

———. *Energy at the Crossroads: Global Perspectives and Uncertainties.*

Strandh, Sigvard. *A History of the Machine.* New York: A. & W. Publishers, 1979.

Vaitheeswarm, Vijay V. *Power to the People: How the Coming Energy Revolution Will Transform an Industry, Change Our Lives, and Maybe Even Save the Planet.* New York: Farrar, Straus & Giroux, 2003.

Wikipedia, "2003 North America Blackout." http://en.wikipedia.org/wiki/ 2003_North_America_Blackout. (Viewed Oct. 30, 2004.)

CREDITS

p. xi: Excerpt from "The Force That Through the Green Fuse Drives the Flower" by Dylan Thomas, from *The Poems of Dylan Thomas*, copyright ©1939 by New Directions Corp., reprinted by permission of New Directions Publishing Corp., and from *Collected Poems* (Dent, 1971), reprinted by permission of David Higham Associates.

p. 19: *Venus of Willendorf*, limestone figure from the Upper Paleolithic, 25th mill. BCE; photograph by Erich Lessing/Art Resource, NY, image #ART29898.

p. 20: Painting of a herd of horses from the Lascaux Caves in Périgord, Dordogne, France; courtesy of Art Resource, NY, image #ART44240.

p. 38: Workers harvesting with sickles, Thebes, Egypt, sixteenth to fourteenth century BCE; photograph by Erich Lessing/Art Resource, NY, image #Art110438.

p. 47: Map of Separate Inventions of Agriculture, from *The Human Web: A Bird's Eye View of World History* by J. R. McNeill and William H. McNeill (New York: W. W. Norton & Company, 2003), p. 27.

p. 49: Ships depicted by Theodor de Bry in *India Orientalis*, 1605; image courtesy of Corbis, image #IH136916.

p. 50: African slaves planting sugar cane under European supervision; Bettmann/Corbis, image # BE036144.

p. 65: Mechanism of a watermill, engraving from V. Zonca's *Nova teatro di machine*, 1607; courtesy of Bibliothèque Nationale de France.

p. 73: Newcomen steam engine, from *Life of Trevithick*, 1775; courtesy of Mary Evans Picture Library, image #10108478.

p. 88: Lamplighter refilling an oil lamp in London—anonymous engraving from Knight's *Old England*, 1800; courtesy of Mary Evans Picture Library, image #10102235.

p. 95: Karl Benz and an assistant seated in the Benz Motorwagon, the first automobile to be sold to the public, 1886; courtesy of Bettman/Corbis, image #SF33082.

p. 110: Buildings of the Chicago World's Fair illuminated at night, photographed by C. D. Arnold, 1893; courtesy of the Chicago Historical Society.

p. 114: Nighttime Map of Our Planet; image by Craig Mayhew and Robert Simmon, NASA GSFC, based on data from the Defense Meteorological Satellite Program, courtesy of NASA/JPL-Caltech.

p. 157: Workers inside the Tokamak Fusion Test Reactor vacuum chamber, in Princeton, New Jersey; photograph by Roger Ressmeyer/Corbis, image #RR003118.

p. 162: Graph of World Energy, 1850–2000, from John Holdren, "Environmental Change and the Human Condition," *American Academy of Arts and Sciences Bulletin*, 57 (Fall 2003), p. 26; © 2004 John P. Holdren.

INDEX

Page numbers in *italics* refer to illustrations.